Agents and Actions Supplements
Vol. 48

Series Editors
K. Brune, Erlangen
M.J. Parnham, Bonn

Prostaglandins and Control of Vascular Smooth Muscle Cell Proliferation

Edited by

K. Schrör
P. Ney

Birkhäuser Verlag
Basel · Boston · Berlin

Editors:

Professor Dr. K. Schrör
Direktor des Institutes für Pharmakologie
Heinrich-Heine-Universität Düsseldorf
Universitätsstr. 1
D-40225 Düsseldorf
Germany

Dr. P. Ney
Abteilung Pharmakologie
Schwarz Pharma AG
Alfred-Nobel-Str. 10
D-40789 Monheim
Germany

A CIP catalogue record for this book is available from the Library of Congress, Washington D.C., USA

Deutsche Bibliothek Cataloging-in-Publication Data
Prostaglandins and control of vascular smooth muscle cell proliferation / ed. by
K. Schrör ; P. Ney. - Basel ; Boston ; Berlin : Birkhäuser, 1997
  (Agents and actions : Supplements ; Vol. 48)
  ISBN-13: 978-3-0348-7354-3   e-ISBN-13: 978-3-0348-7352-9
  DOI: 10.1007/978-3-0348-7352-9
NE: Schrör, Karsten [Hrsg.]; Agents and actions / Supplements

© 1997 Birkhäuser Verlag, P.O. Box 133, CH-4010 Basel, Switzerland
Printed on acid-free paper produced from chlorine-free pulp. TCF ∞
Cover design: Heinz Hiltbrunner, Basel

ISBN-13: 978-3-0348-7354-3

9 8 7 6 5 4 3 2 1

Contents

Preface

Uncontrolled proliferation of vascular smooth muscle cells (SMC) in response to vessel injury is a problem with a considerable therapeutic impact. Specifically, restenosis after percutaneous transluminal coronary angioplasty (PTCA) is a clinical problem without any effective drug therapy so far. Thus, there is need for an improved drug therapy but also for an improved understanding of the pathophysiology of growth control in SMC.

Cyclooxygenase products, such as prostaglandins and thromboxane, are intimately involved in growth responses. Vasodilatory prostaglandins, such as PGI_2, PGE_1 or their analogues, have been shown to inhibit SMC proliferation. There is also evidence for a markedly increased endogenous prostaglandin production during neointima formation under the influence of growth factors which includes induction of COX-2. These data suggest that prostaglandins might be considered both targets and tools of growth control. However, there are still many open questions, including the possible interaction of prostaglandins with other growth-modulating factors, in particular NO, the intracellular signal transduction pathways and the role of oxidative stress.

This book contains a series of invited lectures presented at a Satellite-Symposium „Prostaglandins and Control of Vascular Smooth Muscle Cell Proliferation" which was held on the occasion of the *10th International Conference on Prostaglandins and Related Compounds* in Vienna on September 24th, 1996. It summarizes the state-of-the-art in the field with particular emphasis on prostaglandin-related signal transduction and their possible role in SMC mitogenesis in both experimental and clinical settings. The editors feel that this actual information might be helpful to all physicians and scientists working in this exciting area of vascular biology and medicine.

The success of the Symposium was made possible by generous financial support given by Schwarz Pharma AG (Monheim, Germany). The editors are most grateful to all contributors and participants in the Symposium, Birkhäuser Verlag AG (Basel, Switzerland) for rapid publication of the Proceedings at high technical standards and to Erika Lohmann (Düsseldorf, Germany) for excellent secretarial assistance.

Düsseldorf, January 29th, 1997

Karsten Schrör, M.D.   Peter Ney, Ph.D.

AAS 48
Prostaglandins and Control of Vascular
Smooth Muscle Cell Proliferation
© 1997 Birkhäuser Verlag Basel

Regulation of differentiation/maturation in vascular smooth muscle cells by hormones and growth factors

Gary K. Owens and Gwendolyn Wise

Department of Molecular Physiology and Biological Physics, University of Virginia Health Sciences Center, School of Medicine, P.O. Box 10011, Charlottesville, Virginia, 22906-0011 (USA)

Summary. Smooth muscle cells (SMC) within atherosclerotic lesions show marked alterations in their differentiated properties as compared to normal medial SMC. This process of de-differentiation of SMC has been referred to as "phenotypic modulation", and is characterized by increased growth responsiveness, altered lipid metabolism, increased matrix production, and loss of contractile proteins, all of which can contribute to the development and/or progression of atherosclerotic disease. As such there has been much interest in understanding mechanisms and factors that control the differentiation of the vascular SMC. This paper reviews the effects of growth factors, growth inhibitors, and other extrinsic factors on differentiation/maturation of SMC, with a particular emphasis on consideration of factors that may contribute to abnormal control of SMC differentiation in vascular disease. In addition, we will briefly summarize what is currently known regarding molecular mechanisms that control the coordinate expression of genes encoding for SMC-selective/specific proteins that are required for the differentiated function of the vascular SMC.

Introduction

Intimal migration and proliferation of smooth muscle cells (SMC) are known to play an key role in development of atherosclerotic disease (1,2), and there has been keen interest in identifying factors that have growth promoting and chemotactic activity for vascular SMC, and to determine whether these factors play a role in the atherogenic process. An additional feature of SMC within atherosclerotic lesions is that cells exhibit marked differences in morphology and protein expression patterns as compared to normal medial SMC (3-6). This is characterized by decreased expression of proteins that are characteristic of differentiated smooth muscle (SM) including SM isoforms of contractile proteins, as well as altered growth regulation, lipid metabolism, and decreased contractility [reviewed in (7,8)]. Thus, alterations in the differentiated state of intimal SMC are likely to play a key role in the development and/or the progression of atherosclerosis, and there has been considerable interest in identifying the factors and mechanisms responsible for these changes. Before considering this issue, however, it will first be reviewed what is known regarding control of SMC differentiation under normal conditions.

Properties of differentiated vascular SMC

The primary function of differentiated vascular SMC is contraction, and the cell expresses a unique repertoire of contractile proteins, agonist receptors, ion channels, signaling molecules, etc. to carry out this highly specialized function. Although the principal function of the mature SMC is contraction, the SMC is also capable of a multitude of other functions that vary at different developmental stages, during vascular repair, and in vascular disease [see (7,9) for reviews]. For example, fully differentiated SMC in mature blood vessels proliferate at extremely low rates and produce only small amounts of extracellular matrix proteins. These processes are greatly accelerated during development of the vascular system, during vessel remodeling in hypertension, following vessel injury, and in atherogenesis (7). This plasticity of the SMC is undoubtedly an integral part of the SMC's differentiation program that evolved because it conferred a survival advantage to the organism. For example, when an artery is injured, SMC must be recruited to repair the damage while at the same time maintaining the contractile function of the blood vessel. In this regard, SMC is quite different from cardiac or skeletal muscle which undergo terminal, and essentially irreversible, differentiation and which exhibit much more restricted cellular plasticity (10,11). Control of differentiation in these cell types, while proving to be extremely complex, is nevertheless easier to study because of the stability associated with the terminally differentiated state.

The contractile proteins, and other proteins associated with regulation of contraction, represent logical candidates for use in studying differentiation of the SMC. Mature vascular SMC express unique isoforms of a variety of contractile proteins that are important for their differentiated function. This includes SM α-actin (12,13), SM myosin heavy chains (14,15), SM myosin light chains (16,17), and SM α-tropomyosin (18-20). In addition, differentiated SMC also express a number of proteins that are part of the cytoskeleton and/or purported to be involved in regulation of contraction such as h_1-calponin (21), telokin (22), SM-22α (21), h-caldesmon (23), γ-vinculin (24), (α- and β-) metavinculin (24,25), and desmin (13,26) which show at least some degree of SMC specificity/selectivity [reviewed in (7)]. Expression of the contractile agonist receptors and various signaling molecules must also be considered an integral part of the differentiation program in SMC. However, for the most part these do not appear to be SMC specific/selective and thus may be not useful for identifying cell-type specific regulatory mechanisms.

Expression of the preceding SMC differentiation markers has been shown to be developmentally regulated [reviewed in (7)]. Although no complete study has been reported characterizing the developmental expression of all known SMC differentiation markers in a single species, SM α-actin appears to be the first known marker of differentiated SMC that is expressed during vasculogenesis (26-28). It is first detected in the presumptive SMC that initially envelope the dorsal aortae at stage 12 (day 2 of development) in chicken (27), and quail (28) embryos, or ca. 9 days post-coitus (p.c.) in the developing mouse [Owens, unpublished observations], and rat (29,30). Significantly, at the earliest stages of investment of the aorta with SMC, SM α-actin expression was limited to those presumptive SMC that are in direct proximity to dorsal aortic endothelial cells and was not observed in surrounding mesodermal cells suggesting an important role of endothelial cell-SMC interactions in control of differentiation. Very shortly following detection of SM α-actin there is appearance of additional early SMC differentiation markers including SM22α, SM α-actinin, and α– & β–metavinculin (27,28,31,32). Markers of intermediate stages of vascular SMC differentiation include h-caldesmon, h₁-calponin and the SM1 isoform of SM myosin heavy chain (33-37). The SM2 isoform of SM myosin heavy chain appears to be a late marker of SM differentiation/maturation, appearing in vascular SMC within the aorta predominantly during the postnatal period (33,37).

In summary, the most appropriate differentiation/maturation markers for vascular SMC would appear to be SM α-actin, the SM myosin heavy chains SM-1 and SM-2, h₁-calponin, SM-22 α, telokin, h-caldesmon, gamma-vinculin, and α- and β-metavinculin. Of these, only the SM myosin heavy chains SM1 and SM2 appear to be specific for the SMC lineage (at least to the extent it has been scrutinized thus far), and are capable of identifying SMC to the exclusion of other cell types. It should be noted, however, that complete lineage specificity is not a requirement for a differentiation marker, and an understanding of the mechanisms and factors that control SMC differentiation/maturation will ultimately be dependent on elucidating how the SMC regulates expression of the entire repertoire of genes required for its differentiated function.

Regulation of differentiation/maturation in vascular SMC

General principles

Before considering specific aspects of regulation of differentiation of vascular SMC, it should be first briefly reviewed that a number of general principles control cellular differentiation that have evolved from studies in other cell systems. Cellular differentiation is the process by which multi-potential cells in developing organisms acquire those cell-specific functions and properties that distinguish them from other cell types [see (38) for a review]. Whereas the SMC, like the majority of somatic cells, contains a complete set of genetic material, it expresses only a small number of the genes present. The essence of understanding the control of differentiation is to determine how a cell coordinately regulates the expression of those family of genes necessary for its specialized function. That is: how is it determined which genes will be expressed, when, and at what levels?

The processes whereby multipotential cells acquire those cell-specific characteristics exhibited in mature animals has conventionally been subdivided into three stages, determination, differentiation, and maturation. Determination is the process by which multipotential cells in the developing embryo become committed to a particular cell lineage. Differentiation is the process by which cells that are committed to a particular cell lineage first manifest those cell specific characteristics that distinguish that cell type. Maturation refers to the later stages of differentiation and is characterized by acquisition of further cell specific properties ultimately resulting in the cellular phenotype characteristic of the mature organism. Although these are often considered to be distinct stages, they are really a continuum, with our ability to distinguish the stages being dependent on the extent of our knowledge of the cellular markers that characterize that stage and the molecular processes that control progression from one stage to another. Indeed in most cases the determination event can only be identified retrospectively from cell labeling studies, since by definition cells in this stage have not yet acquired cell specific characteristics that allow them to be recognized. For purposes of simplicity in this review, we will use the term "differentiation" to refer to the entire process by which committed but undifferentiated SMC acquire their cell specific phenotypes, recognizing that the immediate precursor to the differentiated SMC (i.e. a SMC myoblast) and the precise embryological origins of that cell have not, as yet, been clearly identified [see (7) for a review].

Studies in other cell types, principally skeletal muscle, have established a number of general principles of differentiation control that are likely to be applicable to differentiation control in multiple cell systems including vascular SMC. First, it is now known that differentiation involves continuous regulation not simply permanent activation or inactivation of genes, and several families of master regulatory genes have been identified [see reviews by Olson (10), Weintraub (39), and Blau and Baltimore (40)]. For example, in skeletal muscle, it has been shown that MyoD and the related factors myogenin (41,42), myf-5 (43), and MRF/herculin/myf-6 (44), encode transcriptional regulatory factors that are capable of converting a variety of cell lines, including SMC (45), to skeletal myoblasts. The actions of these factors are mediated, at least in part, via the direct activation of a number of muscle specific genes, including muscle creatine kinase, cardiac α-actin, myosin light chain, and troponin I [reviewed by Olson (10) and Weintraub, (39). They are part of a larger family of eukaryotic transcriptional regulators that contain a basic helix loop helix (HLH) motif that is involved in protein dimerization and DNA binding. Members of the MyoD family dimerize with ubiquitously expressed members of the HLH family such as E12, E47, ITF, and subsequently bind a consensus sequence (CANNTG), referred to as an E-box, found in the promoters of many skeletal muscle genes, and activate gene transcription. Additional members of this gene family have been shown to be involved in control of differentiation of a variety of other cell types including adipocytes (46), and neurons (47). The preceding studies have thus established the presence of "master regulatory genes" that are involved in the continuous control of the differentiation program in multiple cell types.

Given that contractile function in vascular SMC is extremely complex and is dependent on a large number of proteins being expressed at the right time and at the appropriate levels, it seems highly likely that differentiation control in SMC will involve at least some "master differentiation" control genes that coordinately regulate expression of the genes required for its differentiated function. However, not all proteins characteristic of differentiated SMC appear to be coordinately regulated, at least at the protein level, either during vascular development, or during phenotypic modulation of SMC in disease (7). As such, if master differentiation control genes exist in SMC, they may be more restrictive in the number of target genes that they effect as compared to the MyoD family, although it is possible that SMC differentiation control genes act on a broad range of target genes but are subject to post-transcriptional controls [see Section B below]. The alternative explanation, that the entire family of genes necessary for differentiated function in SMC are independently regulated, seems highly improbable, particularly in light of recent observations showing the presence of common cis-regulatory elements in the promoters of multiple SMC differentiation marker genes [see references (7,8,48-50) for a review of this area]. However, as yet, no master differentiation control genes have been identified in SMC, nor has a single transcription factor been identified that is either selective or specific for one of the SMC differentiation marker genes.

Extrinsic factors that influence SMC differentiation

The vascular SMC is not terminally differentiated and is capable of undergoing major changes in its phenotype in response to changes in its extracellular milieu. However, the precise nature of environmental cues that affect differentiation of the SMC are relatively poorly understood. A theme that dominated the field for many years was that growth and differentiation were mutually exclusive processes in SMC, and that the differentiated state of the SMC was largely controlled by factors that regulate its growth. However, this is now known to be an over-simplification, and it is appreciated that many factors other than growth state regulate SMC differentiation [reviewed in (7)]. Studies in our laboratory, and others, have demonstrated that the effect of growth stimulation on expression of SMC differentiation markers varies depending on the means of growth stimulation, and the specific SMC differentiation marker examined (14,51-53). For example, whereas PDGF BB markedly suppressed SM α-actin, SM α-tropomyosin, and SM MHC expression, other SMC mitogens including thrombin, basic fibroblast growth factor, fetal bovine serum, epidermal growth factor, or insulin-like growth factor, or combinations of these, did not (7,50,52,53). Effects of PDGF BB were highly selective in that it nearly abolished expression of the SM variants of contractile proteins without affecting the nonmuscle variants. Of interest, PDGF BB-induced decreases in SM α-actin were not associated with changes in transcription, but rather were due to selective mRNA de-stabilization. Moreover, the rate of decrease in SM MHC and SM α-tropomyosin expression following PDGF BB treatment was much more rapid than predicted based on the normal half-life of these mRNAs indicating that mRNA destabilization may also contribute to changes in expression of these SMC differentiation markers, and suggesting that post-transcriptional changes play a major role in the phenotypic modulation of SMC. Additional studies demonstrated: 1) that PDGF BB induced sustained decreases in SM α-actin, SM MHC and SM α–tropomyosin in the absence of sustained mitogenesis; and 2) that PDGF-BB induced suppression of these SMC differentiation marker proteins occurred at concentrations that were 2-4 fold lower than that for mitogenesis (7,52,53). In contrast to the effects of PDGF BB, 10% FBS, which stimulated a mitogenic response equal to or greater than that of PDGF BB, increased, rather than decreased, SM α-tropomyosin expression, and induced only very modest decreases in expression of SM MHCs. In recent studies, we demonstrated that thrombin, which induces a mitogenic response equivalent to that of PDGF

BB (54), stimulated increased, not decreased expression of both SM α-actin and SM MHC (50). Similar evidence for dissociation of the mitogenicity of growth factors and expression of SM MHC, have recently been reported by Somasundaram et al. (55). They found that PDGF AA markedly decreased SM MHC but had no effect on SMC proliferation, whereas $TGF_{\beta 1}$ and IGF-II increased SM MHC expression without effecting cellular growth. Paradoxically these investigators did see some differences regarding effects of specific growth factors as compared to our earlier studies. For example, they found that whereas PDGF BB was a potent mitogen for SMC, it did not decrease SM MHC in their studies, whereas it profoundly reduced it in ours (53). The reason for this difference is unclear but may relate to differences in species (rat versus rabbit aortic SMC), or methods of cell culture. However, results of these studies are in complete agreement in demonstrating that: 1) factors other than growth state per se influence SMC differentiation; and 2) certain "growth factors" may have effects on SMC differentiation that are independent of their effects on cellular growth.

Relatively little is known regarding the effects of growth inhibitors on SMC differentiation. There is extensive evidence showing that agents that elevate cAMP, including prostacyclin (PGI_2), and forskolin, potently inhibit SMC growth [(56-58) , and Fig. 1]. Although to our knowledge no studies have examined effects of these agents on SMC differentiation. We previously demonstrated that $TGF_\beta-$ induced inhibition of the growth of SMC in fetal bovine serum was not associated with significant changes in expression of SM α-actin (52,59) or other SM differentiation marker proteins [Owens and Thompson, unpublished observations]. However, in recent studies we demonstrated that TGF_β stimulated coordinate upregulation of expression of SM α-actin, SM MHC, and h_1-calponin at the mRNA and protein levels, in SMC maintained in a defined serum free media, under conditions where it had no effect on SMC growth [Hautman, Madsen and Owens, manuscript submitted]. Consistent with these observations, Somasundaram et al. (55) demonstrated that $TGF\beta$ increased expression of SM MHC isoforms in cultured rabbit aortic SMC.

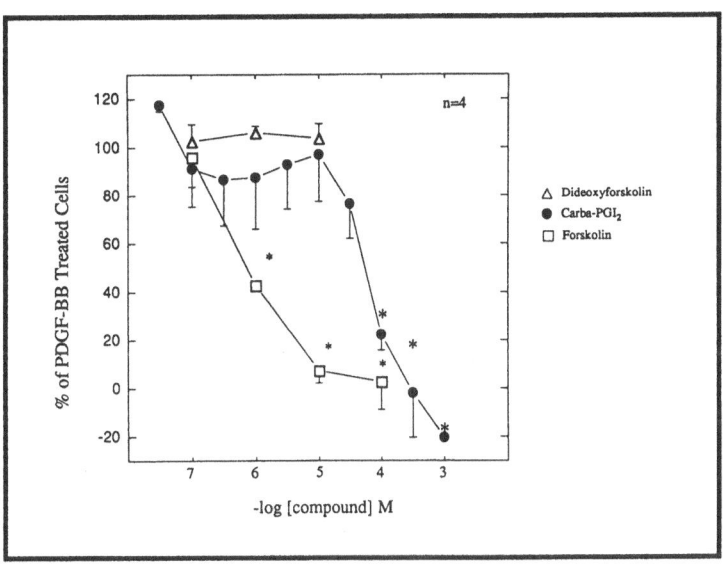

Figure 1. Inhibition of PDGF-BB-induced increases in [³H]-thymidine incorporation by carba-prostacyclin and forskolin. Sub-confluent VSMC were growth-arrested in serum-free medium for 48 hours, and then treated with 20 ng/ml PDGF-BB with or without carba-prostacyclin, forskolin, and 1, 9-dideoxyforskolin. Cells were pulsed with [³H]-thymidine and harvested 24 hours later. The data are expressed as the mean ± s.e.m. * = value significantly less than that obtained in the absence of carba-prostacyclin or forskolin (p < 0.05).

A number of contractile agonists, and neurotransmittors have also been implicated in control of SMC differentiation. We demonstrated that angiotensin II- and vasopressin-induced hypertrophy of cultured SMC was associated with selective increases in expression of multiple SM differentiation marker proteins including SM α-actin, SM tropomyosin, and SM MHC (60). Similarly, Hahn et al. (61) found that chronic treatment of cultured SMC with endothelin inhibited cell cycling, promoted expression of extracellular matrix components, and increased expression of SM α-actin. Loss of innervation has been shown to result in decreased growth and loss of contractility in various smooth muscle tissues (62), and Chamley-Campbell and co-workers (63) reported that the presence of sympathetic neurons delayed phenotypic modulation in primary SMC cultures.

In summary, results of the preceding studies indicate that: 1) a drop in expression of a number of SMC differentiation marker proteins is clearly not obligatory for cell cycle entry in cultured vascular SMC; 2) continued expression of at least some SMC differentiation markers is compatible with proliferation; 3) the level of expression of SM differentiation proteins is not necessarily a direct function of the proliferative activity of the cell but rather appears to be dependent on the specific growth factor or inhibitor involved; and 4) that PDGF BB is unique among the SMC mitogens tested thus far in its ability to selectively suppress the expression of multiple SMC differentiation markers, and that this activity of PDGF does not directly correlate with its proliferative potential (51-53,59,60). Taken together, results suggest that PDGF BB may play an important role in control of SMC differentiation, and that this role may be distinct from its role as a SMC mitogen.

Many other extrinsic factors have also been shown to influence the differentiation/maturation of the SMC. There is extensive evidence showing that mechanical forces can influence growth, matrix production, and differentiation of SMC. Kanda et al. (64) found that mechanical stretch of SMC within a three-dimensional collagen type I matrix increased the abundance of contractile myofilaments in cells, suggesting that mechanical forces promote differentiation of SMC. Birukov et al. (65) demonstrated that cyclic stretch of cultured SMC can exert dual effects on SMC. In high serum containing media, mechanical stretching enhanced SMC growth whereas in low serum containing medium it increased expression of the differentiation marker protein h-caldesmon, but not SM myosin heavy chain (MHC) or h_1-calponin. Ives and co-workers (66,67) found that passive stretch of cultured SMC using a FlexerCell unit increased expression of SM MHC isoforms in SMC maintained in a defined serum free media. Interestingly, these workers also demonstrated that cyclic stretching increased autocrine production of PDGF AA and PDGF BB (66), which are potent negative differentiation factors for SMC (52,53), and that administration of anti-PDGF antibodies enhanced the effects of stretch on SM MHC expression. Mechanical stretching has also been shown to influence expression of extracellular matrix components by SMC (68), and to increase expression of the α_1 and α_2- subunits of the Na^+/K^+-ATPase (69). Cheng et al. (70) found that transient compression of SMC on a collagen gel stimulated a delayed increase in DNA synthesis which was nearly completely blocked with a neutralizing antibody to FGF-2. In summary, results indicate that the effects of mechanical forces on SMC are complex, involving stimulation of growth factors, and/or negative differentiation factors in some cases, but promotion of SMC differentiation under other circumstances. Clearly further studies are needed to determine which response occurs in vivo under physiological versus pathological conditions.

Extracellular matrix components are likely to play a key role in control of differentiation of vascular SMC (71,72). Work from many laboratories has shown the presence of multiple extracellular matrix components in vasculogenic regions in developing embryos (73,74). Moreover, Drake et al. (75) demonstrated that administration of beta-1 integrin neutralizing antibodies arrested assembly of angioblasts (i.e. endothelial cell precursors) into the primordial aorta in the developing chick embryo. The specific matrix components and integrins that mediate differentiation of SMC in vivo during vasculogenesis have not yet been identified. There is, however, clear evidence based on studies in cultured SMC, showing that extracellular matrix components influence SMC differentiation. For example, Thyberg and Nilsson (76) demonstrated that growth of cultured SMC on laminin and collagen type IV increased expression of SM α-actin whereas growth on fibronectin decreased it, suggesting that these matrix components have differential effects on SMC differentiation. Growth of cultured SMC on Matrigel, a basement-membrane rich matrix material isolated from an EHS tumor, has also been shown to increase expression of SM α-actin, and contractile agonist-induced calcium responses as compared to cells grown on standard tissue culture plastic (77).

There is extensive circumstantial evidence showing that endothelial cells play a key role in differentiation of SMC during vascular development (78). All blood vessels initially exist as endothelial tubes which become invested with smooth muscle cells during formation of larger vessels including arteries and arterioles [a process referred to as arterialization]. However, the mechanisms whereby endothelial cells influence SMC differentiation, and the factors that determine whether a given vessel undergoes arterialization have not been clearly identified [reviewed in (79)]. Endothelial cells are known to produce a large number of factors that influence smooth muscle growth and/or differentiation including PDGF, bFGF, connective tissue growth factor, nitric oxide, PGI_2, and many others (80). Chamley-Campbell et al. (81) reported that endothelial cell conditioned medium, or co-culture of primary cultures of SMC with confluent, but not subconfluent endothelial cells, inhibited growth and prevented or delayed at least some of the phenotypic changes that occur in SMC when placed in culture. Addition of heparin to the culture media had similar effects, raising the possibility that effects may be mediated by heparin or heparin-like compounds. Both cultured endothelial cells and SMC are known to secrete a heparin-like molecule that inhibits SMC growth (82,83). Heparin has also been shown to inhibit myointimal proliferation of SMCs following vascular injury in vivo (84,85), although it did not prevent injury-induced decreases in SM α-actin expression. Heparin has been reported to increase SM α–actin expression in cultured SMC, but only under conditions in which it had an antiproliferative effect (86), suggesting that effects were secondary to growth inhibition rather than a direct effect on SMC differentiation per se. In contrast to the preceding observations showing endothelial cell induced inhibition of SMC growth, many studies have shown growth stimulatory effects of endothelial cells (9,87). It is well established that endothelial cells express PDGF which as discussed above is both a potent mitogen and negative regulator of SMC differentiation (51,52). Endothelial cells also produce a variety of other SMC mitogens including connective tissue growth factor, and bFGF (80,88). In addition, there is extensive evidence showing that endothelial cell co-culture, as well as endothelial cell conditioned media promotes rather than inhibits SMC growth (9,87). Consistent with these observations, we recently demonstrated that conditioned media from rat aortic endothelial cells stimulated rather than inhibited growth of rat aortic SMC , and potently and selectively suppressed expression of SM α-actin, SM myosin heavy chains, and SM α-tropomyosin (89). This activity was sensitive to heat and protease treatment, was not inhibited by neutralizing antibodies to PDGF or bFGF, bound weakly to heparin sepharose, and had an estimated molecular size of 45 kDa based on gel filtration. The identity of this factor (or factors) has not yet been established. There is thus evidence that endothelial cells secrete both positive and negative SMC differentiation factors. However, relatively little is known regarding what regulates the balance between these opposing activities. Barrett et al. (90) demonstrated that endothelial cells in culture produce much higher levels PDGF transcripts than do their in vivo counterparts (90). One possibility is that the effects of endothelial cells on SMC differentiation vary as a function of the growth/differentiated state of the endothelial cell. Key questions include: Which EC-derived factors regulate the differentiation of vascular SMC in vivo ? What regulates EC production of both positive and negative SMC differentiation factors? Do SMC influence the differentiated state of the endothelial cell? What is the effect of atherogenic stimuli on expression of endothelial factors that effect SMC differentiation?

Effects of atherogenic stimuli on SMC differentiation in atherosclerosis

Whereas the precise nature of the initiating event for development of atherosclerosis is not known, it is clear that the failure of the SMC to maintain its normal differentiated phenotype within the atherosclerotic lesion is likely to be a key contributing factor in the progression of the disease. However, as yet, few studies have examined the effects of specific atherogenic substances such as lipids or lipid peroxidation products on SMC differentiation per se. It is also interesting to consider how the response of the SMC to atherogenic stimuli may have evolved. In this regard, it is important to remember that atherosclerosis is a disease that rarely manifests its effects until late in life - i.e. beyond the normal reproductive years. Moreover, whereas man has evolved over millions of years, atherosclerosis has only been a major health problem recently as the average lifespan of mankind increased to the point where individuals died of it as opposed to other causes. Many atherosclerotic risk factors related to diet, lifestyle, etc. are also the consequences of modern civilization. The net consequence is that there has not been, nor is there likely to be, significant genetic selection pressure against the disease. As such, from a teological point of view, the nature of the response of the SMC to atherosclerotic stimuli presumably evolved as a consequence of other selection pressures that affect survival of the organism prior to sexual maturity. The most logical candidate is the ability of the SMC to carry out vascular repair, since it is clear that impairment of this property would have major negative consequences for the survival capabilities of the organism. From this perspective, atherogenic stimuli may be viewed of as a relatively recent challenge to the system. The SMC responds using mechanisms that evolved to a different form of injury, such as mechnical trauma. The net consequence is that the outcome is not ideal - some responses are beneficial (e.g. formation of a stable fibrous cap), whereas others are not (lumenal encroachment). Thus, whereas atherosclerosis has been present in the human species for a very long time, from an evolutionary standpoint it represents a relatively new challenge to our systems.

Conclusions/Future directions

From the preceding, it is evident that differentiation/maturation of vascular SMC is highly dependent on a complex array of environmental cues (Figure 2), and that this property confers a high degree of cellular plasticity to the cell, enabling it to undergo rapid changes in phenotype to meet the physiological demands of the organism. Although much progress has been made in recent years, much remains to be learned regarding the mechanisms and factors that control the normal differentiation of smooth muscle cells during vascular development, as well as the alterations in the differentiated state of SMC characteristic in vascular diseases such as atherosclerosis. A particularly important issue is to identify the molecular mechanisms that control transcription of genes that are required for the differentatiated function of the SMC, and to determine how environmental factors known to influence SMC differentiation alter these control processes. It is clear that our understanding of SMC differentiation/maturation is in its infancy and that much remains to be learned regarding all aspects of this process. However, we are entering a very exciting time in which there is likely to be rapid progress. Only when we better understand how SMC differentiation/maturation is normally regulated, are we likely to understand how these control processes are altered in atherosclerosis, and how the resulting changes in the phenotype of the SMC contribute to lesion development and progression.

Acknowlegdements
Supported by Grants P01 HL19242 and R01 HL38854 from the National Institutes of Health (USA).

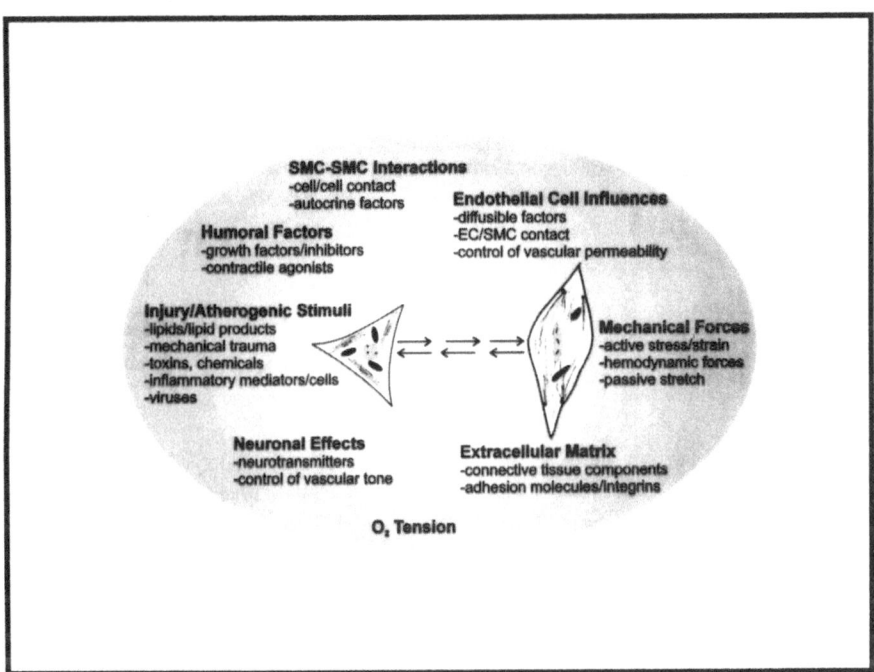

Figure 2. Summary of extrinsic factors or local environmental cues believed to be important in controlling the differentiation/maturation of the vascular smooth muscle cell. This figure illustrates that differentiation/maturation of SMC is dependent on the complex interaction of multiple local environmental signals, not any single factor, and that modification of any one of these factors may contribute to an alteration in the differentiated state of the SMC (i.e. phenotypic modulation). The SMC can exhibit a wide range of different phenotypes ranging from the highly synthetic proliferative cell on the **left** to the mature, fully contractile cell on the **right**. The multiple arrows connecting the two SMC phenotypes are meant to illustrate the multiplicity of phenotypes available between the two extremes shown, and that the changes appear to be reversible. Figure reprinted from Owens (96) with permission from the publisher.

References

1.   Schwartz SM, Ross R. Cellular proliferation in atherosclerosis and hypertension. Progr Cardiovasc Disease 1984; 26:355-372.

2.   Ross R, Glomset JA. The pathogenesis of atherosclerosis. N Engl J Med 1976; 295:369-377/420-425.

3.   Kocher O, Gabbiani F, Gabbiani G, Reidy MA, Cokay MS, Peters H, Huttner I. Phenotypic features of smooth muscle cells during the evolution of experimental carotid artery intimal thickening. Biochemical and morphologic studies. Lab Invest 1991; 65:459-470.

4.   Kocher O, Gabbiani G. Cytoskeletal features of normal and atheromatous human arterial smooth muscle cells. Hum Pathol 1986; 17:875-880.

5.   Mosse PR, Campbell GR, Campbell JH. Smooth muscle phenotypic expression in human carotid arteries. II. Atherosclerosis-free diffuse intimal thickenings compared with the media. Arteriosclerosis 1986; 6:664-669.

6.   Glukhova MA, Kabakov AE, Frid MG, Ornatsky OI, Belkin AN, Mukhin DN, Orekhov AN, Koteliansky VE, Smirnov VN. Modulation of human aorta smooth muscle cell phenotype: a study of muscle-specific variants of vinculin, caldesmon, and actin expression. Proc Natl Acad Sci USA 1988; 85:9542-9546.

7.   Owens GK. Regulation of differentiation of vascular smooth muscle cells. Physiol Rev 1995; 75:487-517.

8.   Owens GK. Role of alterations in the differentiated state of vascular SMC in atherogenesis. In: Atherosclerosis and Coronary Artery Disease. Ross R, Fuster V, Topol E, editors. New York: Raven Press, 1995; 401-420.

9.   Schwartz SM, Campbell GR, Campbell JH. Replication of smooth muscle cells in vascular disease. [Review]. Circ Res 1986; 58:427-444.

10.   Olson EN. Regulation of muscle transcription by the MyoD family. The heart of the matter. Circ Res 1993; 72:1-6.

11.   Blau H, Pavlath G, Hardeman E, et al. Plasticity of the differentiated state. Science 1985; 230:758-766.

12.   Owens GK, Thompson MM. Developmental changes in isoactin expression in rat aortic smooth muscle cells in vivo. Relationship between growth and cytodifferentiation. J Biol Chem 1986; 261:13373-13380.

13.   Gabbiani G, Schmid E, Winter S, Chaponnier C, de Ckhastonay C, Vandekerckhove J, Weber K, Franke WW. Vascular smooth muscle cells differ from other smooth muscle cells: predominance of vimentin filaments and a specific alpha-type actin. Proc Natl Acad Sci USA 1981; 78:298-302.

14.   Rovner AS, Murphy RA, Owens GK. Expression of smooth muscle and nonmuscle myosin heavy chains in cultured vascular smooth muscle cells. J Biol Chem 1986; 261:14740-14745.

15.   Rovner AS, Thompson MM, Murphy RA. Two different heavy chains are found in smooth muscle myosin. Am J Physiol 1986; 250:C861-C870.

16.   Hasegawa Y, Ueda Y, Watanabe M, Morita F. Studies on amino acid sequences of two isoforms of 17-kDa essential light chain of smooth muscle myosin from porcine aorta media. J Biochem 1992; 111:798-803.

17.   Helper DJ, Lash JA, Hathaway DR. Distribution of isoelectric variants of the 17,000-
       dalton myosin light chain in mammalian smooth muscle. J Biol Chem 1988; 263:15748-
       15753.

18.   Bretscher A. Thin filament regulatory proteins of smooth muscle and nonmuscle cells.
       Nature 1986; 321:726-727.

19.   Marston SB, Smith CJW. The thin filaments of smooth muscle. J Muscle Res Cell
       Motility 1985; 6:669-708.

20.   Lees-Miller JP, Helfman DM. The molecular basis for tropomyosin isoform diversity.
       Bioessays 1991; 13:429-437.

21.   Winder SJ, Sutherland C, Walsh MP. Biochemical and functional characterization of
       smooth muscle calponin. [Review]. Adv Exp Med Biol 1991; 304:37-51.

22.   Herring BP, Smith AF. Telokin expression is mediated by a smooth muscle cell-specific
       promoter. Am J Physiol 1996; 270:C1656-C1665.

23.   Sobue K, Sellers JR. Caldesmon, a novel regulatory protein in smooth muscle and
       nonmuscle actomyosin systems. J Biol Chem 1991; 266:12115-12118.

24.   Geiger B, Tokuyasu KT, Dutton AH, Singer SJ. Vinculin, an intracellular protein
       localized at specialized sites where microfilament bundles terminate at cell membranes.
       Proc Natl Acad Sci USA 1980; 77:4127-4131.

25.   Pardo JV, Siliciano JD, Craig SW. Vinculin is a component of an extensive network of
       myofibril-sarcolemma attachment regions in cardiac muscle fibers. J Cell Biol 1983;
       97:1081-1088.

26.   Mitchell JJ, Reynolds SE, Leslie KO, Low RB, Woodcock-Mitchell J. Smooth muscle
       cell markers in developing rat lung. Am J Respir Cell Mol Biol 1990; 3:515-523.

27.   Duband JL, Gimona M, Scatena M, Sartore S, Small JV. Calponin and SM 22 as
       differentiation markers of smooth muscle: spatiotemporal distribution during avian
       embryonic development. Differentiation 1993; 55:1-11.

28.   Hungerford JE, Owens GK, Argraves WS, Little CD. Development of the aortic vessel
       wall as defined by vascular smooth muscle and extracellular matris markers. Devel Biol
       1996; 178:375-392.

29.   Gomez AR, Sturgill BC, Chevalier RL, Boyd DG, Lessard JL, Owens GK, Peach MJ.
       Fetal expression of muscle-specific isoactins in multiple organs of the Wistar-Kyoto rat.
       Cell Tissue Res 1987; 250:7-12.

30.   Sawtell NM, Lessard JL. Cellular distribution of smooth muscle actins during
       mammalian embryogenesis: expression of the alpha-vascular but not the gamma-enteric
       isoform in differentiating striated myocytes. J Cell Biol 1989; 109:2929-2937.

31.   Li L, Miano JM, Cserjesi P, Olson EN. SM22 alpha, a marker of adult smooth muscle, is expressed in multiple myogenic lineages during embryogenesis. Circ Res 1996; 78:188-195.

32.   Samaha FF, Ip HS, Morrisey EE, Seltzer J, Tang Z, Solway J, Parmacek MS. Developmental pattern of expression and genomic organization of the calponin-h1 gene. A contractile smooth muscle cell marker. J Biol Chem 1996; 271:395-403.

33.   Miano J, Cserjesi P, Ligon K, Perisamy M, Olson EN. Smooth muscle myosin heavy chain marks exclusively the smooth muscle lineage during mouse embryogenesis. Circ Res 1994; 75:803-812.

34.   Sartore S, Scatena M, Chiavegato A, Faggin E, Giuriato L, Pauletto P. Myosin isoform expression in smooth muscle cells during physiological and pathological vascular remodeling. J Vasc Res 1994; 31:61-81.

35.   Kuro-o M, Nagai R, Tsuchimochi H, Katoh H, Yazaki Y, Ohkubo A, Takaku F. Developmentally regulated expression of vascular smooth muscle myosin heavy chain isoforms. J Biol Chem 1989; 264:18272-18275.

36.   Frid MG, Shekhonin BV, Koteliansky VE, Glukhova MA. Phenotypic changes of human smooth muscle cells during development: late expression of heavy caldesmon and calponin. Dev Biol 1992; 153:185-193.

37.   Aikawa M, Sivam PN, Kuro-o M, Kimura K, Nakahara K, Takewaki S, Ueda M, Yamaguchi H, Yazaki Y, Periasamy M. Human smooth muscle myosin heavy chain isoforms as molecular markers for vascular development and atherosclerosis. Circ Res 1993; 73:1000-1012.

38.   Darnell J, Lodish H, Baltimore D. Molecular Cell Biology. New York: W.H. Freeman and Company, 1986; 987-1035.

39.   Weintraub H, Davis R, Tapscott S, Thayer M, Krause M, Benezra R, Blackwell TK, Turner D, Rupp R, Hollenberg S. The myoD gene family: nodal point during specification of the muscle cell lineage. Science 1991; 251:761-766.

40.   Blau H, Baltimore D. Differentiation requires continuous regulation. J Cell Biol 1991; 112:781-783.

41.   Wright WE, Sassoon DA, Lin VK. Myogenin, a factor regulating myogenesis, has a domain homologous to MyoD. Cell 1989; 56:607-617.

42.   Edmondson DG, Olson EN. A gene with homology to the myc similarity region of MyoD1 is expressed during myogenesis and is sufficient to activate the muscle differentiation program. Genes Dev 1989; 3:628-640.

43. Braun T, Buschhausen-Denker G, Bober E, Tannich E, Arnold HH. A novel human muscle factor related to but distinct from MyoD1 induces myogenic conversion in 10T1/2 fibroblasts. EMBO J 1989; 8:701-709.

44. Rhodes SJ, Konieczny SF. Identification of MRF4: a new member of the muscle regulatory factor gene family. Genes Dev 1989; 3:2050-2061.

45. van Neck JW, Medina JJ, Onnekink C, van der Ven PF, Bloemers HP, Schwartz SM. Basic fibroblast growth factor has a differential effect on MyoD conversion of cultured aortic smooth muscle cells from newborn and adult rats. Am J Pathol 1993; 143:269-282.

46. Tontonoz P, Kim JB, Graves RA, Spiegelman BM. ADD1: a novel helix-loop-helix transcription factor associated with adipocyte determination and differentiation. Mol Cell Biol 1993; 13:4753-4759.

47. Lee JE, Hollenberg SM, Snider L, Turner DL, Lipnick N, Weintraub H. Conversion of Xenopus ectoderm into neurons by neuroD, a basic helix-loop-helix protein. Science 1995; 268:836-844.

48. Shimizu RT, Blank RS, Jervis R, Lawrenz-Smith SC, Owens GK. The smooth muscle α-actin gene promoter is differentially regulated in smooth muscle versus non-smooth muscle cells. J Biol Chem 1995; 270:7631-7643.

49. Madsen CS, Hershey JC, Owens GK. Identification of a positive cis element in the rat smooth muscle myosin heavy chain promoter. FASEB J 1996; 10:A343 [abstract].

50. Owens GK, Vernon SM, Madsen CS. Molecular regulation of smooth muscle cell differentiation. J Hypertens 1996; 14 (suppl 5):S56-S64.

51. Corjay MH, Thompson MM, Lynch KR, Owens GK. Differential effect of platelet-derived growth factor- versus serum-induced growth on smooth muscle alpha-actin and nonmuscle beta-actin mRNA expression in cultured rat aortic smooth muscle cells. J Biol Chem 1989; 264:10501-10506.

52. Blank RS, Owens GK. Platelet-derived growth factor regulates actin isoform expression and growth state in cultured rat aortic smooth muscle cells. J Cell Physiol 1990; 142:635-642.

53. Holycross BJ, Blank RS, Thompson MM, Peach MJ, Owens GK. Platelet-derived growth factor-BB-induced suppression of smooth muscle cell differentiation. Circ Res 1992; 71:1525-1532.

54. McNamara CA, Sarembock IJ, Gimple LW, Fenton JW, Coughlin SR, Owens GK. Thrombin stimulates proliferation of cultured rat aortic smooth muscle cells by a proteolytically activated receptor. J Clin Invest 1993; 91:94-98.

55.   Muhlhauser J, Merrill MJ, Pili R, Maeda H, Becic M, Bewig B, Passaniti A, Edwards NA, Crystal RG, Capagrossi MC. VEGF165 expressed by a replication-deficient recombinant adenovirus vector induces angiogenesis in vivo. Circ Res 1995; 77:1077-1086.

56.   Grosser T, Bönisch D, Zucker TP, Schrör K. Iloprost-induced inhibition of proliferation of coronary artery smooth muscle cells is abolished by homologous desensitization. Agents Actions 1995; 45(suppl):85-91.

57.   Hara S, Morishita R, Tone Y, Yokoyama C, Inoue H, Kaneda Y, Ogihara T, Tanabe T. Overexpression of prostacyclin synthase inhibits growth of vascular smooth muscle cells. Biochem Biophys Res Commun 1995; 216:862-867.

58.   Koh E, Morimoto S, Jiang B, Inoue T, Nabat T, Kitano S, Yasuda O, Fukuo K, Ogihara T. Effects of beraprost sodium, a stable analogue of prostacyclin, on hyperplasia, hypertrophy and glycosaminoglycan synthesis of rat aortic smooth muscle cells. Artery 1993; 20:242-252.

59.   Blank RS, Thompson MM, Owens GK. Cell cycle versus density dependence of smooth muscle alpha actin expression in cultured rat aortic smooth muscle cells. J Cell Biol 1988; 107:299-306.

60.   Turla MB, Thompson MM, Corjay MH, Owens GK. Mechanisms of angiotensin II- and arginine vasopressin-induced increases in protein synthesis and content in cultured rat aortic smooth muscle cells. Evidence for selective increases in smooth muscle isoactin expression. Circ Res 1991; 68:288-299.

61.   Hahn AW, Resink TJ, Kern F, Buhler FR. Effects of endothelin-1 on vascular smooth muscle cell phenotypic differentiation. J Cardiovasc Pharmacol 1992; 20(suppl 12):S33-36.

62.   Bevan RD, Tsuru H. Functional and structural changes in the rabbit ear artery after sympathetic denervation. Circ Res 1981; 49:478-485.

63.   Chamley-Campbell J, Campbell GR, Ross R. The smooth muscle cell in culture. Physiol Rev 1979; 59:1-6.

64.   Kanda K, Matsuda T. Mechanical stress-induced orientation and ultrastructural change of smooth muscle cells cultured in three-dimensional collagen lattices. Cell Transplant 1994; 3:481-492.

65.   Birukov KG, Shirinsky VP, Stepanova OV, Tkachuk VA, Hahn AW, Resink TJ, Smirnov VN. Stretch affects phenotype and proliferation of vascular smooth muscle cells. Mol Cell Biochem 1995; 144:131-139.

66.   Wilson E, Mai Q, Sudhir K, Weiss RH, Ives HE. Mechanical strain induces growth of vascular smooth muscle cells via autocrine action of PDGF. J Cell Biol 1993; 123:741-747.

67. Reusch HP, Wagdy H, Reusch R, Wilson E, Ives HE. Mechanical strain increases smooth muscle and decreases nonmuscle myosin expression in rat vascular smooth muscle cells. Circ Res 1996; 79:1046-1053.

68. Leung D, Glagov S, Mathews M. A new in vitro method for studying cell responses to mechanical stimulation: different effects of cyclic stretching and agitation on smooth muscle cell biosynthesis. Exp Cell Res 1977; 109:285-298.

69. Songu-Mize E, Liu X, Stones JE, Hymel LJ. Regulation of Na^+,K^+-ATPase alpha-subunit expression by mechanical strain in aortic smooth muscle cells. Hypertension 1996; 27:827-832.

70. Cheng GC, Libby P, Grodzinsky AJ, Lee RT. Induction of DNA synthesis by a single transient mechanical stimulus of human vascular smooth muscle cells. Role of fibroblast growth factor-2. Circulation 1996; 93:99-105.

71. Hedin U, Sjolund M, Hultgardh-Nilsson A, Thyberg J. Changes in expression and organization of smooth-muscle-specific alpha-actin during fibronectin-mediated modulation of arterial smooth muscle cell phenotype. Differentiation 1990; 44:222-231.

72. Pauly RR, Passaniti A, Crow M, Kinsella JL, Papadopoulos N, Monticone R, Lakatta EG, Martin GR. Experimental models that mimic the differentiation and dedifferentiation of vascular cells. Circulation 1992; 86:III68-III73.

73. Carey DJ. Control of growth and differentiation of vascular cells by extracellular matrix proteins. Annu Rev Physiol 1991; 53:161-177.

74. Little CD, Piquet DM, Davis LA, Walters L, Drake CJ. Distribution of laminin, collagen type IV, collagen type I, and fibronectin in chicken cardiac jelly/basement membrane. Anat Rec 1989; 224:417-425.

75. Drake CJ, Davis LA, Little CD. Antibodies to beta 1-integrins cause alterations of aortic vasculogenesis, in vivo. Devel Dyn 1992; 193:83-91.

76. Thyberg J, Hultgardh-Nilsson A. Fibronectin and the basement membrane components laminin and collagen type IV influence the phenotypic properties of subcultured rat aortic smooth muscle cells differently. Cell Tissue Res 1994; 276:263-271.

77. Li X, Tsai P, Wieder ED, Kribben A, Van Putten V, Schrier RW, Nemenoff RA. Vascular smooth muscle cells grown on matrigel. A model of the contractile phenotype with decreased activation of mitogen-activated protein kinase. J Biol Chem 1994; 269:19653-19658.

78. Poole TJ, Coffin JD. Vasculogenesis and angiogenesis: two distinct morphogenetic mechanisms establish embryonic vascular pattern. J Exp Zool 1989; 251:224-231.

79. Skalak TC, Price RJ. The role of mechanical stresses in microvascular remodeling. Microcirculation 1996; 3:143-165.

80.   Grotendorst GR. Growth factors as regulators of wound repair. [Review]. Int J Tissue React 1988; 10:337-344.

81.   Campbell JH, Campbell GR. Endothelial cell influences on vascular smooth muscle phenotype [review]. Annu Rev Physiol 1986; 48:295-306.

82.   Castellot JJ, Favreau LV, Karnovsky MJ, Rosenberg RD. Inhibition of vascular smooth muscle cell growth by endothelial cell-derived heparin. Possible role of a platelet endoglycosidase. J Biol Chem 1982; 257:11256-11260.

83.   Fritze LM, Reilly CF, Rosenberg RD. An antiproliferative heparan sulfate species produced by postconfluent smooth muscle cells. J Cell Biol 1985; 100:1041-1049.

84.   Clowes AW, Clowes MM, Fingerle J, Reidy MA. Regulation of smooth muscle cell growth in injured artery [review]. J Cardiovasc Pharmacol 1989; 14 (suppl 6):S12-S15.

85.   Clowes AW, Clowes MM, Kocher O, Ropraz P, Chaponnier C, Gabbiani G. Arterial smooth muscle cells in vivo: Relationship between actin isoform expression and mitogenesis and their modulation by heparin. J Cell Biol 1988; 107:1939-1945.

86.   Desmouliere A, Rubbia Brandt L, Gabbiani G. Modulation of actin isoform expression in cultured arterial smooth muscle cells by heparin and culture conditions. Arterioscler Thromb 1991; 11:244-253.

87.   Gajdusek CM, Schwartz SM. Ability of endothelial cells to condition culture medium. J Cell Physiol 1982; 110:35-42.

88.   Bradham DM, Igarashi A, Potter RL, Grotendorst GR. Connective tissue growth factor: a cysteine-rich mitogen secreted by human vascular endothelial cells is related to the SRC-induced immediate early gene product CEF-10. J Cell Biol 1991; 114:1285-1294.

89.   Vernon SM, Campos MJ, Haystead T, Thompson MM, Dicorleto PE, Owens GK. Endothelial cell conditioned media downregulates smooth muscle contractile protein expression. Am J Physiol 1996; 272: (in press).

90.   Barrett TB, Gajdusek CM, Schwartz SM, McDougall JK, Benditt EP. Expression of the sis gene by endothelial cells in culture and in vivo. Proc Natl Acad Sci USA 1984; 81:6772-6774.

AAS 48
Prostaglandins and Control of Vascular
Smooth Muscle Cell Proliferation
© 1997 Birkhäuser Verlag Basel

Novel indices of oxidant stress in cardiovascular disease: specific analysis of F₂-isoprostanes

Domenico Praticò, Murdeach Reilly, John A. Lawson and Garret A. FitzGerald*

*The Center for Experimental Therapeutics, University of Pennsylvania, Philadelphia, PA, USA.
905 Stellar Chance Laboratories, University of Pennsylvania, 422 Curie Blvd., Philadelphia, PA 19104. Tel
215-898-7056; Fax 215-573-8996; email garret@spirit.gcrc.upenn.edu

Summary. The development of methods to measure specific isoprostanes affords a unique opportunity to investigate both the role of oxidant stress as a mechanism of disease *in vivo* and to select rational doses of putative antioxidant drugs and vitamins for evaluation in human disease. The ability to measure these compounds directly *in situ* at the site of their formation, to immunolocalize them to target cells in atherosclerotic plaque and other tissues (61) and to assess their biosynthesis non-invasively in urine promises to elucidate the role of lipid peroxidation in cardiovascular disease.

Introduction

Modification of proteins, lipids and DNA by free radicals-oxidant stress-has been implicated as a mechanism of pathogenesis in diseases as disparate as atherosclerosis and cancer (1,2). However, our ability to obtain evidence in direct support of this hypothesis has been limited by the paucity of methods which permit investigation of the process *in vivo*. Several constraints apply to conventional approaches to the study of reactive oxygen species (ROS) generation *in vivo*. Commonly, the targets for analysis are chemically unstable and/or susceptible to generation *ex vivo*. This applies to the measurement of circulating lipid peroxides and conjugated dienes (3,4). Alternatively, reliance is placed on indices of peroxidation or free radical generation explicitly induced *ex vivo*. Examples of this approach include the generation of dienes during copper induced peroxidation of low density lipoprotein (LDL) *ex vivo* (5,6) and the generation of spin traps *ex vivo*, as detected by electron spin resonance (ESR) (7). However, little information is available to facilitate interpretation of how changes in these indices may relate to oxidant stress *in vivo*. For example, how does the oxidizability of LDL relate to actual LDL oxidation *in vivo*?

Given the kinetics of their generation and the reactivity of free radicals themselves, it would seem unreasonable, presently, to measure them directly in biological fluids. However, an indirect assay of ROS generation might, ideally, have several features. Firstly, it would target a chemically stable analyte, which was formed *in vivo*, specifically by a free radical catalyzed process. Secondly, the susceptibility to formation of this analyte *ex vivo*, after sample collection, would be absent. Thirdly, the assay method would be specific, sensitive and precise. Fourthly, the analyte might also be formed *in situ*, at the site of free radical generation, as well as in biological fluids, which might be sampled. Finally, such an assay would be most useful if it were not inordinately labor intensive or expensive to perform.

Presently, no assays satisfy all of these requirements. Consequently, our knowledge of how ROS generation is associated with human disease, never mind its functional importance, is extremely limited. Similarly, we have very little information on the dose-response relationships and comparative potencies of antioxidant vitamins and drugs (8,9). Despite this, consumption of vitamins for their antioxidant properties is widespread amongst the "worried well" in

Western societies and support for phase three clinical trials of these compounds has generated confusion, rather than enlightenment (10,11).

Several novel approaches to the study of oxidant stress *in vivo* have been described. These include assays developed towards products of free radical attack on DNA or DNA adduct formation by lipid peroxides *in vivo* (12,13). Additionally, novel products of lipid peroxidation have been identified and are being evaluated as indices of oxidant stress. These include the isoprostanes (14), the subject of this review.

The isoprostanes

These compounds are free radical catalyzed products of arachidonic acid. They are isomers of the conventional, enzymatically formed eicosanoids. Presently, they include members of the F, D and E series, isothromboxanes and isoleukotrienes (15, 16, 17, 18). It is likely that they will encompass a parallel array of eicosanoid-like structures and that analogous 1 and 3 series compounds may be formed from appropriate substrates. Several groups had demonstrated the formation of prostaglandin-like auto-oxidation products from arachidonic acid *in vitro* (19,20); however, it was the observation of Morrow and Roberts that F_2-isoprostanes were formed initially *in situ* in the phospholipid and were then cleaved out to circulate in plasma, which has prompted investigation of these compounds as potential indices of oxidant stress *in vivo* (21,22). Attention has focussed upon the F_2 series and 8-epi $PGF_{2\alpha}$ in particular, due to the biological activity of the latter compound (*vide infra*).

Several approaches to investigation of the isoprostanes have been described. Morrow and Roberts estimated F_2-isoprostane formation using a gas chromatography mass/spectrometry (GC/MS) based, selected ion monitoring assay which employed an isotope labeled $PGF_{2\alpha}$ internal standard. They have commonly applied this assay to measurements in plasma, using conventional extraction techniques to estimate the isoprostane content circulating freely vs that esterified in lipid (23). Such an approach has provided much useful information, which has recently been summarized (24). However, it also has intrinsic limitations. Thus, despite the sensitivity of the GC/MS approach, the specificity and precision of the method is constrained by the absence of an isoprostane internal standard, given that $PGF_{2\alpha}$ elutes at some distance from the F_2 isoprostanes. Secondly, the presence of circulating phospholipids rendery the measurement susceptible to confounding by isoprostane formation *ex vivo*, when the technique is applied to plasma. A similar constraint applies to the use of plasma thromboxane (Tx) B_2 as an index of TxA_2 biosynthesis (25). Our experience in this instance suggested that the problem could be bypassed by measurement of an *in vivo* TxB_2 metabolite in plasma (26). A major metabolite of 8-epi $PGF_{2\alpha}$ has recently been identified (27) and may allow refinement of the current approach to estimating formation of this compound in plasma.

Isoprostanes are also excreted in urine. We reasoned that this might be preferable to plasma for initial isoprostane analysis. Samples could be obtained noninvasively and the relative absence of phospholipid minimized the chance of *ex vivo* formation. Experience with conventional prostaglandin analysis suggested that, assay limitations aside, the information afforded by this approach would be likely to be at least as useful in interpreting alterations in isoprostane biosynthesis as that provided by measurements of the same compounds in plasma (25, 27, 28). We also wished to obtain specific information on one compound, which we could then extend to the wider array of isoprostanes, as appropriate. Given its properties as a vasoconstrictor and platelet active agent (29-32), we focussed initially upon 8-epi $PGF_{2\alpha}$.

Urinary 8-epi $PGF_{2\alpha}$; an index of oxidant stress in vivo

Initially, we developed a selected ion monitoring assay for the compound, using an $[^{18}O_2]$-labeled internal standard. Excretion proved to be remarkably constant in healthy individuals. There was no evident circadian variation and moderate exercise did not alter excretion of the isoprostane. Although we did not observe differences in excretion between genders, levels tended to increase at the extremes of age.

To address the hypothesis that it might be an index of oxidant stress, we first examined the pattern of 8-epi $PGF_{2\alpha}$ excretion in patients with syndromes putatively associated with excessive free radical generation. For example, poisoning with paracetemol or paraquat, both of which have been suggested to result in oxidant stress were associated with increased 8-epi PGF$_{2\alpha}$ excretion, which fell as the symptoms and signs of poisoning resolved (33). Similarly, cigarette smoke is a source of many radicals (34,35). Additionally, smoking may result in cellular activation - for example, platelets and neutrophils (36,37) - which may, in turn, be a source of free radical generation. Others have demonstrated that apparently healthy individuals who smoke cigarettes had increased excretion of 8-oxo-7,8 dihydro-2'-deoxyguanosine (8-oxodG), an index of free radical modification of DNA (38). Morrow and colleagues (39) reported that total urinary isoprostanes were elevated in response to smoking. Similarly, we showed that smoking resulted in a dose dependent increase in excretion of 8-epi PGF$_{2\alpha}$. Furthermore, these elevated levels fell when the volunteers quit smoking and switched to nicotine patches (40).

Free radical generation has been implicated in reperfusion injury (41-44). We have studied isoprostane biosynthesis in a variety of human models of coronary reperfusion (45). For example, excretion is increased in patients undergoing coronary artery bypass grafting, peaking at the period coinciding with clamp release. Similarly, urinary excretion is increased in patients presenting with myocardial infarction who are treated with either rt-PA or by angioplasty. Interestingly, urinary 8-epi PGF$_{2\alpha}$ is not increased above that observed in age and gender matched controls in patients with stable coronary artery disease. However, experience in several patients in whom samples were obtained prior to administration of thrombolytics is consistent with an increase in excretion corresponding to the period of vascular reperfusion. We have biochemically simulated this condition in a canine model of coronary reperfusion, where excretion of the isoprostane may be more readily related to changes in coronary flow (45).

There are limited data which relate 8-epi PGF$_{2\alpha}$ excretion to other indices of oxidant stress. Excretion is closely related in time to detection of an adduct of PBN by electron spin resonance in 2 patients who underwent coronary reperfusion with rt-PA (45). Both coronary artery bypass graft (CABG) patients and cigarette smokers are potential human models of oxidant stress in whom this issue might be addressed. Similarly, we and others are in the initial stages of gathering information which might clarify the dose-response relationships of vitamins and antioxidant drugs on biosynthesis of isoprostanes. The ability to measure this compound in mouse urine (Figure 1) affords the opportunity to study oxidant stress during the evolution of vascular disease in transgenic models of atherosclerosis and related diseases.

Enzymatic formation of 8-epi PGF$_{2\alpha}$

Examination of the formation of isoprostanes during platelet activation was surprising. It appeared that 8-epi PGF$_{2\alpha}$, but not other isoprostanes, was formed by an aspirin sensitive mechanism (46). This compound, alone of the F$_2$ isoprostanes, could be formed by the semi-purified platelet cyclooxygenase (COX-1). The identity of the compound was confirmed as 8-epi PGF$_{2\alpha}$ by a variety of approaches, including electron impact mass spectrometry and sequential chromatographic purifications (46). Although the amount of 8-epi PGF$_{2\alpha}$ generated was far less than the amounts of either thromboxane B$_2$ or 12-HETE formed during platelet activation, it was produced coincidentally with these products. Furthermore, all three were suppressed by aspirin or indomethacin (46).

These observations were consistent with those of Hecker et al.(47) who added radiolabelled arachidonic acid to the semipurified COX and demonstrated a peak of radioactivity consistent with formation of 8-epi PGF$_{2\alpha}$ as a minor enzymatic product. They represented both an interesting observation and a potential problem. Obviously, formation of 8-epi PGF$_{2\alpha}$ as a prostaglandin might undermine its utility as an index of oxidant stress *in vivo*. Several of the syndromes of oxidant stress which we had studied - cigarette smoking and coronary reperfusion - are associated with platelet activation *in vivo*, as reflected by excretion of urinary metabolites of thromboxane (48-50).

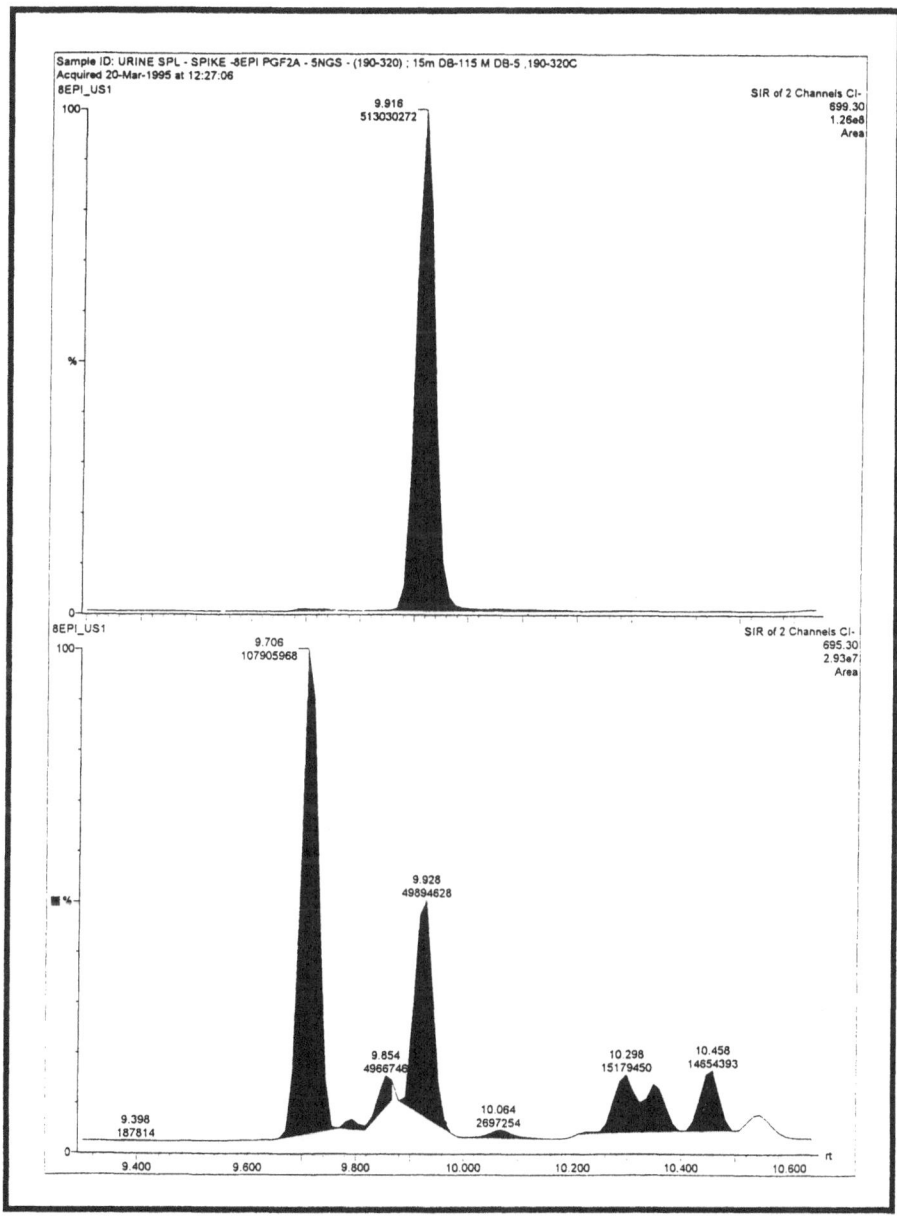

Figure 1. Detection of 8-epi PGF$_{2\alpha}$ in mouse urine. The upper trace depicts the internal standard and the lower trace the endogenous material.

To address this issue, we studied chronic cigarette smokers. We investigated the effects of a low dose aspirin regimen on comparative suppression of 8-epi $PGF_{2\alpha}$ and thromboxane formation.

Low dose aspirin, as expected, completely suppressed platelet COX-1 activity, as reflected by serum TxB_2 (Figure 2). 8-epi $PGF_{2\alpha}$ was formed in serum and was suppressed by aspirin. However, the suppression was incomplete, averaging 80%. Thus, human platelets had the capacity to form both compounds *ex vivo* via an aspirin sensitive pathway; what of actual biosynthesis? Again, as expected (51), the aspirin regimen suppressed urinary excretion of a thromboxane metabolite by about 75%. This is consistent with platelets being the major, albeit not the only, source of thromboxane biosynthesis. Excretion was increased in the smokers, as we have previously reported, consistent with platelet activation *in vivo*. The effect of aspirin on 8-epi $PGF_{2\alpha}$ excretion was quite different. Despite its partial inhibition of platelet capacity to form the compound, actual biosynthesis, as reflected by its excretion in urine, was not altered by aspirin. These results suggested that the enzymatic pathway contributed trivially, if at all, to overall biosynthesis of 8-epi $PGF_{2\alpha}$, even in a syndrome of COX-1 activation.

To address the hypothesis that the more readily inducible COX-2 might form 8-epi $PGF_{2\alpha}$, we turned to monocytes. These cells express both forms of the enzyme. We pretreated the cells with aspirin to inhibit the constitutive COX-1 and then induced COX-2 with bacterial lipopolysaccharide (LPS) (52). Again, 8-epi $PGF_{2\alpha}$ formation was observed coincident with expression of the enzyme and formation of the two conventional prostaglandin products, PGE_2 and TxB_2. Suppression of all products, including 8-epi $PGF_{2\alpha}$, was achieved with a selective inhibitor of COX-2. Steroids, which down regulated the enzyme, cycloheximide and actinomycin D had similar effects. Interestingly, stimulation of these cells in a distinct experimental setting illustrated their capacity to form 8-epi $PGF_{2\alpha}$ in a free radical dependent manner. Thus, zymosan stimulation of monocytes in the presence of low density lipoprotein (LDL) results in formation of 8-epi $PGF_{2\alpha}$ which is suppressed by the antioxidants BHT and SOD, but not by the selective inhibitor of COX-2. Formation of PGE_2 and TxB_2 is not observed in these stimulated cells (52).

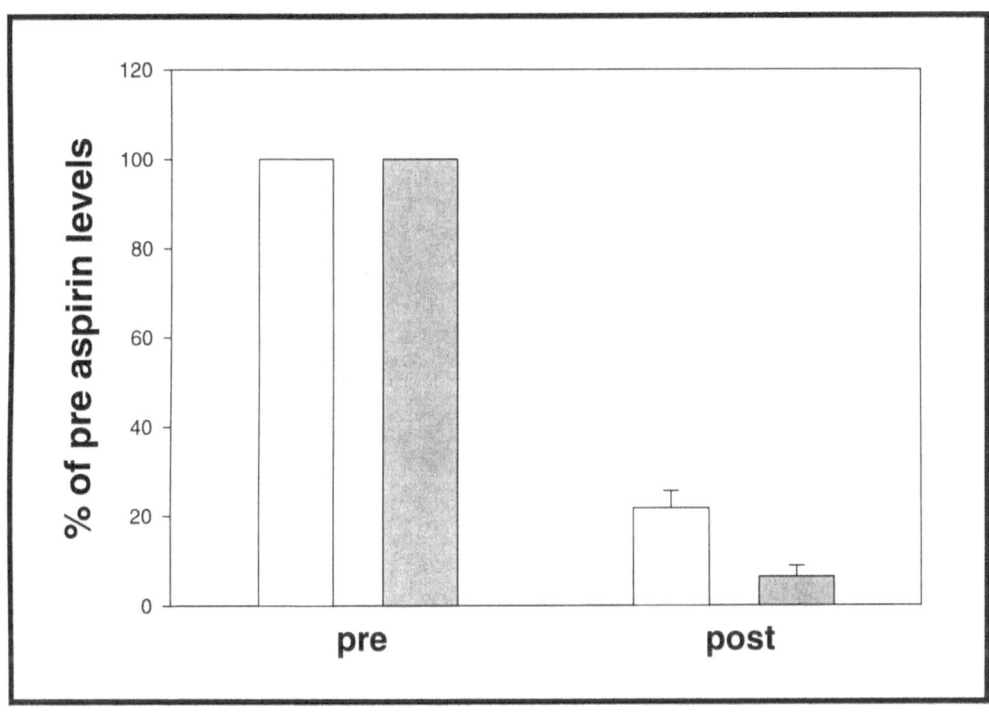

Figure 2. The percentage change in serum 8-epi PGF$_{2\alpha}$ (open columns) and TxB$_2$ (solid columns) in smokers (n=4) treated with aspirin (75 mg/d for 10 days). Suppression of both compounds by aspirin was significant (P<0.01).

In summary, 8-epi PGF$_{2\alpha}$ can be formed as a trivial product of either COX in human cells. However, the evidence to date suggests that the contribution of this pathway to overall biosynthesis of the compound is trivial in healthy volunteers and in at least one syndrome of COX-1 activation. Given the need to exclude the contribution of this pathway in the multiple clinical situations (e.g.; inflammation, ischemia-reperfusion, cancer) where COX activation and free radical generation might coincide, it seemed desirable to develop methods targeted on other isoprostanes not formed by the COX pathways.

Specific estimation of distinct isoprostanes in human disease

Four theoretical classes of F_2 isoprostanes (Figure 3) may be formed (53). While 8-epi $PGF_{2\alpha}$ is a member of the class IV and thus may be termed IP (isoprostane)F_2-IV, we have recently synthesized an internal standard for an IPF_2-I (54). Development of an assay for this compound allowed us determine that it was even more abundant in human urine than 8-epi $PGF_{2\alpha}$. Chronic administration of low dose aspirin and acute administration of high dose indomethacin failed to suppress urinary $IPF_{2\alpha}$-I (Lawson, Barry, Praticò, Rokach and FitzGerald unpublished, 1996). We have recently reported coordinate measurement of both IPFs in human syndromes of oxidant stress.

Oxidative modification of LDL is thought of importance in the process of atherogenesis (55-57). We (45) and others (58, 59) have demonstrated that oxidation of LDL by copper or co-incubation with endothelial cells results in rapid formation of 8-epi $PGF_{2\alpha}$, coincident with generation of lipid peroxides. We have reported elevated excretion of urinary 8-epi $PGF_{2\alpha}$ in patients with homozygous type 2 hypercholesterolemia, compared with age and gender matched controls. Heterozygotes also exhibit a significant increase in biosynthesis of the isoprostane. However, the increment is not as pronounced as in the homozygotes. Interestingly, we have also noted that the LDL content (unstimulated) of 8-epi $PGF_{2\alpha}$ in the latter group is also elevated compared to that in their controls (60). Measurement of IPF_2-I in both groups demonstrated that excretion was also elevated in hypercholesterolemia and that excretion of the two isoprostanes was highly correlated, both in the homozygotes ($r=0.78$; $p < 0.0001$) and in the heterozygotes ($r=0.65$; $p < 0.0001$).We have also observed highly correlated increments in excretion of both isoprostanes in pregnancy induced hypertension and in patients presenting with myocardial infarction undergoing PTCA.

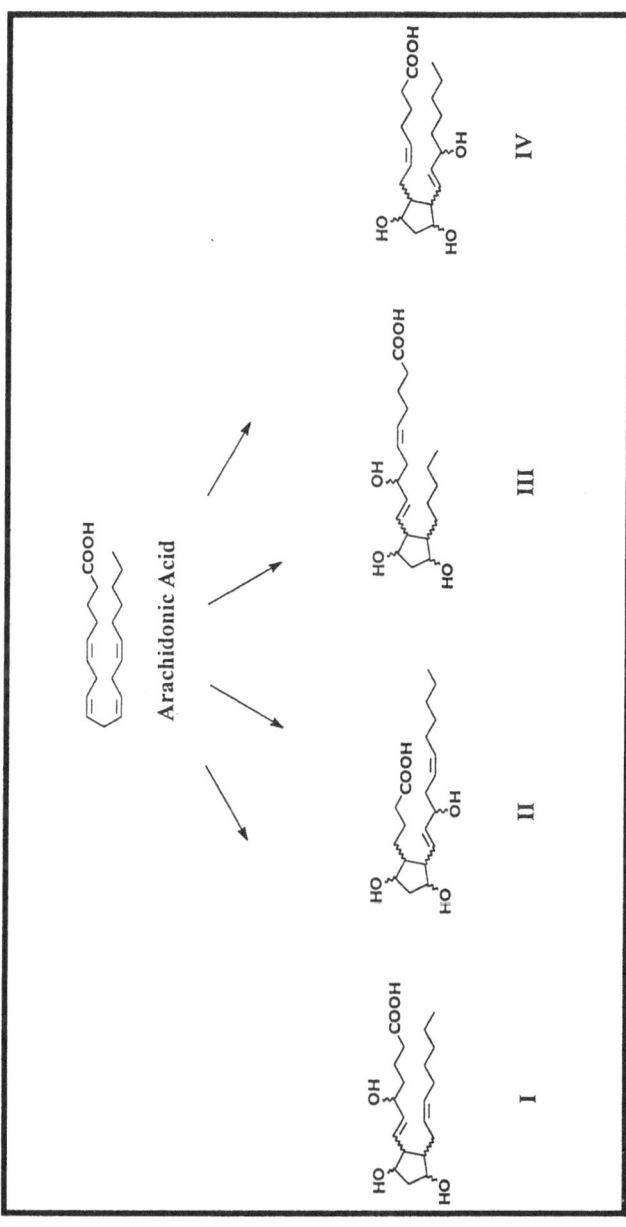

Figure 3. The four types of F_2 isoprostanes originating from arachidonic acid.

These observations, made in highly disparate human syndromes of oxidant stress, suggest that the two isoprostanes are being formed by a common mechanism--in this case free radical attack on arachidonic acid. They are consistent with our experiments using aspirin, in that they suggest that COX-dependent 8-epi $PGF_{2\alpha}$ formation is likely to contribute trivially, if at all, to overall systemic biosynthesis of the compound, as reflected by its excretion in urine.

Acknowledgements
This work was supported by grants from the Wellcome Trust and the NIH (HL 56000 and M01RR0040). Dr. FitzGerald is the Robinette Foundation Professor of Cardiovascular Medicine. Dr. Reilly held a fellowship from the Irish Heart Foundation during the course of this work.

References

1.   Halliwell B, Gutteridge JMC. Free Radical in Biology and Medicine. (2nd edition), Oxford University Press, Oxford, United Kingdom,1989.

2.   Gutteridge JMG. Free radicals in disease processes: a compilation of cause and consequence. Free Radic Res Commun 1993; 19:141-158.

3.   Holley AE, Slater TF. Measurement of lipid hydroperoxydes in normal human blood plasma using HPLC-chemiluminescence linked to a diode array detector for measuring conjugated dienes. Free Radic Res Commun 1991; 15:51-63.

4.   Gutteridge JMC. Aspects to consider when detecting and measuring lipid peroxidation. Free Radic Res Commun 1986; 1:173-184.

5.   Dujovne CA, Harris WS, Colle Gerrond LL, Fan J, Muzio F. Comparison of effects of probucol versus vitamin E on *ex vivo* oxidation susceptibility of lipoproteins in hyperlipoproteinemia. Am J Cardiol 1994; 74:38-42.

6.   Lenz ML, Hughes H, Mitchell JR, Via DP, Guyton JR, Taylor AA, Gotto AMJr, Smith CV. Lipid hydroperoxy and hydroxy derivatives in copper-catalyzed oxidation of low density lipoprotein. J Lipid Res 1990; 31:1043-1050.

7.   Janzen EG. Spin trapping and associated vocabulary. Free Radic Res Commun 1990; 10:63-68.

8.   Kleinveld HA, Demacker PNM, Stalenhoef AFH. Comparative study on the effect of low-dose vitamin E and probucol on the susceptibility of LDL to oxidation and the progression of atherosclerosis in Watanabe Heritable Hyperlipemic rabbits. Arterioscl Thromb Vasc Biol 1994; 14:1386-1391.

9.   Kleinveld HA, Kak-Lemmers HLM, Hectors MPC, De Fouw NJ, Demacker PNM, Stalenhoef AFH. Vitamin E and fatty acid intervention does not attenuate the progression of atherosclerosis in Watanabe Heritable Hyperlipemic rabbits. Arterioscl Thromb Vasc Biol 1995; 15:290-297.

10.   Steinberg D. Antioxidants and atherosclerosis. Circulation 1991; 84:1420-1425.

11.   Packer L. Protective role of vitamin E in biological systems. Am J Clin Nutr 1991; 53:1050-1055.

12.   Shigenaga MK, Aboujaoude EN, Chen Q, Ames BN. Assays of oxidative DNA damage biomarkers 8-oxo-2'-deoxyguanosine and 8-oxoguanine in nuclear DNA and biological fluids by high-performance liquid chromatography with electrochemical detection. Methods Enzymol 1994; 234:16-33.

13.   Adachi S, Kawamura K, Takemoto K. Oxidative damage of nuclear DNA in liver of rats exposed to psychological stress. Cancer Res 1993; 53:4153-4155.

14.   Morrow JD, Harris TM, Roberts LJ. Non-cyclooxygenase oxidative formation of a series of novel prostaglandins: analytical ramifications for measurement of eicosanoids. Anal Biochem 1990; 84:1-10.

15.   Morrow JD, Hill HE, Burk RF, Nammour TM, Badr KF, Roberts LJ. A series of prostaglandin F_2 like compounds are produced in vivo in humans by a non-cyclooxygenase free radical catalyzed mechanism. Proc Natl Acad Sci USA 1990; 87:9383-9387.

16.   Morrow JD, Minton TA, Mukundan CR, Campbell MD, Zackert WE, Daniel VC, Badr KF, Blair IA, Roberts LJ. Free radical-induced generation of isoprostanes in vivo: Evidence for the formation of D-ring and E-ring isoprostanes. J Biol Chem 1994; 269:4317-4326.

17.   Morrow JD, Awad JA, Wu A, Zackert WE, Daniel VC, Roberts LJ. Nonenzymatic free radical-catalyzed generation of thromboxane-like compounds (isothromboxanes) in vivo. J Biol Chem 1996; 271:23185-23190.

18.   Harrison KA, Murphy RC. Isoleukotrienes are biologically active free radical products of lipid peroxidation. J Biol Chem 1995; 270:7273-17278.

19.   Pryor WA, Stanley JP, Blair E. Autoxidation of polyunsaturated fatty acids: II. A suggested mechanism for the formation of TBA-reactive materials from prostaglandin-like endoperoxides. Lipids 1976; 11:370-379.

20.   O'Connor DE, Mihelich ED, Coleman MC. Stereochemical course of the autoxidative cyclization of lipid hydroperoxides to prostglandin-like bicyclo endoperoxides. J Am Chem Soc 1984; 106:3577-3584.

21. Morrow JD, Awad JA, Boss HJ, Blair IA, Roberts LJ. Non-cyclooxygenase-derived prostanoids (F_2-isoprostanes) are formed in situ on phospholipids. Proc Natl Acad Sci USA 1992; 89:10721-10725.

22. Morrow JD, Awad JA, Kato T, Takahashi K, Badr KF, Roberts LJ, Burk RF. Formation of novel non-cyclooxygenase-derived prostanoids (F_2-isoprostanes) in carbon tetrachloride hepatoxicity. An animal model of lipid peroxidation. J Clin Invest 1992; 90:2502-2507.

23. Morrow JD, Roberts LJ. Mass spectrometry of prostanoids: F_2-isoprostanes produced by non-cyclooxygenase free radical-catalyzed mechanism. Methods Enzymol 1994; 233:163-1174.

24. Morrow JD, Roberts LJ. The isoprostanes. Current knowledge and directions for future research. Biochem Pharmacol 1996; 51:1-9.

25. FitzGerald GA, Pedersen AK, Patrono C. Analysis of prostacyclin and thromboxane A_2 biosynthesis in cardiovascular disease (editorial). Circulation 1983; 67:1174-1177.

26. Catella F, Healy D, Lawson J, FitzGerald GA. ll-dehydro-thromboxane B_2: An index of thromboxane formation in the human circulation. Proc Natl Acad Sci USA 1986; 83: 5861-5865.

27. Roberts LJ, Moore KP, Zackert WE, Oates JA, Morrow JD. Identification of the major urinary metabolites of the F_2-isoprostane 8-isoprostaglandin $F_{2\alpha}$ in humans. J Biol Chem 1996; 271:20617-20620.

28. Kuehl PG, Bolds JM, Lloyd JE, Snapper J, FitzGerald GA. Thromboxane A_2/prostaglandin endoperoxide activation mediates bronchial and hemodynamic responses to endotoxemia in the conscious sheep. Am J Physiol 1988; 254: (Regulatory Integrative Comp Physiol 23:) R310-R319.

29. Takahashi K, Nammour TM, Fukunaga M, Ebert J, Morrow JD, Roberts LJ, Hoover RL, Badr KF. Glomerular actions of a free radical-generated novel prostaglandin, 8-epi-prostaglandin $F_{2\alpha}$, in the rat. Evidence for interaction with thromboxane A_2 receptors. J Clin Invest 1992; 90:136-141.

30. Banerjee M, Kang KH, Morrow JD, Roberts LJ, Newman JH. Effects of a novel prostaglandin, 8-epi-$PGF_{2\alpha}$, in rabbit lung in situ. Am J Physiol 1992; 32:H660-H663.

31. Yin K, Halushka PV, Yan Y-T, Wong PY-K. Antiaggregatory activity of 8-Epi-Prostaglandin $F_{2\alpha}$ and other F-series prostanoids and their binding to Thromboxane A_2/Prostaglandin H_2 receptors in human platelets. J Pharmacol Exp Ther 1994; 270:1192-1196.

32. Praticò D, Smyth EM, Violi F, FitzGerald GA. Local amplification of platelet function by 8-epi-prostaglandin $F_{2\alpha}$ is not mediated by thromboxane receptor isoforms. J Biol Chem 1996; 271:14916-12924.

33. Delanty N, Reilly M, Praticò D, Fitzgerald DJ, Lawson JA, FitzGerald GA. 8-epi-$PGF_{2\alpha}$: specific analysis of an isoeicosanoid as an index of oxidant stress in vivo. Br J Clin Pharmacol 1996; 42:15-19.

34. Pryor WA, Stone K. Oxidants in cigarette smoke: radicals, hydrogen peroxide, peroxynitrate and peroxynitrite. Ann NY Acad Sci 1993; 686:12-27.

35. Kalra J, Chaudhary AK, Prasad K. Increased production of oxygen free radicals in cigarette smokers. Int J Exp Pathol 1991; 72:1-7.

36. Nowak J, Murray JJ, Oates JA, FitzGerald GA. Biochemical evidence of a chronic abnormality in platelet and vascular function in healthy individuals who smoke cigarettes. Circulation 1987; 76:6-14.

37. Lehr HA, Kress E, Menger MD, Frield HP, Hubner C, Arfors KE, Messmer KE. Cigarette smoke elicits leukocyte adhesion to endothelium in hamsters: inhibition by CuZn-SOD. Free Radic Biol Med 1993; 14:573-581.

38. Loft S, Astrup A, Buemann B, Poulsen HE. Oxidative DNA damage correlates with oxygen consumption in humans. FASEB J 1994; 8:534-537.

39. Morrow JD, Frei B, Longmire AW, Gaziano JM, Lynch SM, Shyr Y, Strauss WE, Oates JA, Roberts LJ. Increase in circulating products of lipid peroxidation (F_2-isoprostanes) in smokers. Smoking as a cause of oxidative damage. N Engl J Med 1995; 332:1198-1203.

40. Reilly M, Delanty N, Lawson JA, FitzGerald GA. Modulation of oxidant stress in vivo in chronic cigarette smokers. Circulation 1996; 94:19-25.

41. Davies SW, Ranjadayalan K, Wickens DG, Dormandy TL, Timmis AD. Lipid peroxidation associated with successful thrombolysis. Lancet 1990; 335:741-741.

42. Purvis J, Young I, Lightbody J, Trimble E, Adgey J. Free radical production following therapy in acute myocardial infarction. Circulation 1992; 86:I-804 [abstract].

43. Kloner RA, Przykklenk K, Whittaker P. Deleterious effects of oxygen radicals in ischemia/reperfusion. Resolved and unresolved issues. Circulation 1989; 80:1115-1127.

44. Loesser KE, Kukreja RC, Kazziha N, Jesse RL, Hess ML. Oxidative damage to the myocardium: a fundamental mechanism of myocardial injury. Cardioscience 1991; 2:199-216.

45.   Delanty N, Reilly M, Praticò D, Lawson JA, McCarthy JF, Wood AE, Ohnishi ST, Fitzgerald DJ, FitzGerald GA. 8-epi-PGF$_{2\alpha}$ generation during coronary reperfusion: a potential quantitative marker of oxidant stress in vivo. Circulation 1997 (in press)

46.   Praticò D, Lawson JA, FitzGerald GA. Cyclooxygenase-dependent formation of the isoprostane, 8-epi-Prostaglandin F$_{2\alpha}$. J Biol Chem 1995; 270:9800-9808.

47.   Hecker M, Ullrich V, Fischer C, Meese CO. Identification of novel arachidonic acid metabolites formed by prostaglandin H synthase. Eur J Biochem 1987; 169:113-123.

48.   Fitzgerald DJ, FitzGerald GA. Role of thrombin and thromboxane A$_2$ in reocclusion following thrombolysis with tissue-type plasminogen activator. Proc Natl Acad Sci USA 1989; 86:7585-7589.

49.   Kerins DM, Roy L, FitzGerald GA, Fitzgerald DJ. Platelet and vascular function during coronary thrombolysis with tissue-type plasminogen activator. Circulation 1989; 80:1718-1725.

50.   Fitzgerald DJ, Wright F, FitzGerald GA. Increased thromboxane biosynthesis during coronary thrombolysis. Evidence that platelet activation and thromboxane A$_2$ modulate the response to tissue-type plasminogen activator in vivo. Circ Res 1989; 65:83-94.

51.   Catella F, FitzGerald GA. Paired analysis of urinary thromboxane B$_2$ metabolites in humans. Thromb Res 1987; 47:647-656.

52.   Praticò D, FitzGerald GA. Generation of 8-epi-Prostaglandin F$_{2\alpha}$ by human monocytes. Discriminate production by reactive oxygen species and prostaglandin endoperoxide synthase-2. J Biol Chem 1996; 271:8919-8924.

53.   Waugh RJ, Murphy RC. Mass spectrometric analysis of four regioisomers of F$_2$-isoprostanes formed by free radical oxidation of arachidonic acid. J Am Soc Mass Spectrom 1996; 7:490-499.

54.   Adiyaman M, Lawson JA, Hwang SW, Khanapure SP, FitzGerald GA, Rokach J. Total synthesis of a novel isoprostane IP F$_{2\alpha}$ and its identification in biological fluids. Tetrahedron Lett 1996; 37:4849-4852.

55.   Steinberg D, Parthasarathy S, Carew TE, Khoo JC, Witztum JL. Beyond cholesterol: modification of low-density lipoprotein that increase its atherogenicity. N Engl J Med 1989; 320:915-924.

56.   Witztum JL, Steinberg D. Role of oxidized low density lipoprotein in atherogenesis. J Clin Invest 1991; 88:1785-1792.

57.   Holvoet P, Collen D. Oxidized lipoproteins in atherosclerosis and thrombosis. FASEB J 1994; 8:1279-1284.

58.   Lynch SM, Morrow JD, Roberts LJ, Frei B. Formation of non-cyclooxygenase-derived prostanoids (F_2-isoprostanes) in plasma and low density lipoprotein exposed to oxidative stress in vitro. J Clin Invest 1994; 93:998-1004.

59.   Gopaul NK, Zadeh-Nourooz J, Mallet AI, Anggard EE. Formation of F_2-isoprostanes during aortic endothelial cell-mediated oxidation of low density lipoprotein. FEBS Lett 1994; 348:297-300.

60.   Reilly M, Delanty N, Tremoli E, Rader D, FitzGerald GA. Elevated levels of 8-epi-Prostaglandin $F_{2\alpha}$ in familial hypercholesterolemia: evidence for oxidative stress in vivo. Circulation 1996; 94:3727 A.

61.   Praticò D, Iuliano L, Spagnoli L, Mauriello A, Maclouf J, Violi F, FitzGerald GA. Monocytes in human atherosclerotic plaque contain high levels of 8-epi-$PGF_{2\alpha}$: an index of oxidative stress in vivo. Circulation 1996; 94:1611 A.

AAS 48
Prostaglandins and Control of Vascular
Smooth Muscle Cell Proliferation
© 1997 Birkhäuser Verlag Basel

Role of thromboxane A$_2$ in mitogenesis of vascular smooth muscle cells

Gerald W. Dorn II

Division of Cardiology, University of Cincinnati and the Cincinnati VA Medical Cente, 231 Bethesda Avenue, ML 042, Cincinnati, Ohio 45267-0542, U.S.A., Phone: (513)558-4619; Fax: (513)558-3116; E-Mail: DornGW@UCBEH.SAN.UC.EDU

Summary. Thromboxane A$_2$, a product of activated platelets, is a potent vasoconstrictor and promoter of vascular smooth muscle cell growth. Therefore, thromboxane has the potential to contribute to processes, such as restenosis following coronary angioplasty, characterized by both platelet activation and abnormal vascular smooth muscle growth. This article reviews the effects of thromboxane on growth of cultured vascular smooth muscle cells, discusses the mechanisms by which thromboxane transduces its growth promoting effects in tissue culture with an emphasis on the role of endogenously produced basic fibroblast growth factor, and reviews clinical studies of thromboxane synthesis inhibitors and/or receptor blockers in angioplasty restenosis.

Introduction

Vascular smooth muscle migration and proliferation with formation of an obstructive neointima results after vascular injury associated with angioplasty or atherectomy, in experimental and clinical atherosclerosis, in systemic and pulmonary hypertension, and in the accelerated atherosclerosis which occurs after orthotopic heart transplantation (1-5). These types of vascular injury are typified by physical or chemical endothelial injury followed by platelet activation with release of platelet derived growth promoting substances. Animal studies have shown that experimental thrombocytopenia attenuates, but does not prevent neointima formation after balloon injury (6) and recent studies suggest a similar beneficial effect of platelet inhibition with anti Gp IIb/IIIa antibodies on angioplasty restenosis (7). Although a great deal of attention has been given to the effects of platelet-derived peptide growth factors such as PDGF in initiating vascular smooth muscle growth, it is becoming increasingly clear that other platelet-derived substances can also modify vascular growth. Foremost among these is thromboxane A$_2$ which is formed and released by activated platelets and which has been demonstrated to have independent growth promoting effects, as well as synergistic effects with other growth factors, in cultured vascular smooth muscle cells.

In this paper we outline our present understanding of the intracellular pathways regulating thromboxane-stimulated growth of vascular smooth muscle cells, and discuss the implications of this phenomenon in coronary artery restenosis after percutaneous rcvascularization procedures. For purposes of this review, studies employing stable thromboxane agonists will be described as "thromboxane-stimulated ...", even though authentic thromboxane was not employed due to its chemical instability. Also, vascular smooth muscle "growth" is a general term which will be used to describe either hypertrophy or hyperplasia or both. When a specific growth response is alluded to, it will be defined as such.

Vascular smooth muscle growth in arterial disease

In normal arteries the muscular wall or media consists of quiescent, myosin rich vascular smooth muscle cells which contract when exposed to vasoconstricting agonists, but which are generally unresponsive to growth factors (8). During *in vitro* tissue culture, vascular smooth muscle cells lose the ability to contract as the capacity to proliferate is forced upon them by the tissue culture environment. The modulated phenotype of cultured vascular smooth muscle cells is characterized by relatively less contractile actin and myosin protein expression and increased expression of secreted matrix proteins such as collagen and thrombospondin. Interestingly, these are also characteristics of activated vascular smooth muscle cells which form vascular neointima after balloon injury in animals and balloon angioplasty in humans.

Abnormal vascular smooth muscle growth, either hyperplastic or hypertrophic, has been noted in several disease states. In large blood vessels of chronically hypertensive patients and animals smooth muscle growth is primarily due to enlargement of previously existing cells (hypertrophy), without cell proliferation (9,10). In contrast, vascular growth after induction of acute, experimental hypertension is typically a combination of migration and proliferation (11). As noted above, vascular growth after balloon injury appears to be proliferative, with additional lumenal compromise secondary to increased synthesis of extracellular matrix proteins. Cultured vascular smooth muscle cells can also be stimulated to hypertrophy or proliferate, depending upon the nature of the stimulus, the source of the cells, and the culture conditions (see below).

There is accumulating evidence that regulated expression of endogenous growth modifiers, both positive and negative, is an important means by which vascular smooth muscle growth responses are controlled. Antibodies to basic fibroblast growth factor (bFGF) and platelet-derived growth factor (PDGF) attenuate neointima formation after balloon injury in animal models (12-14). Dzau and colleagues have postulated that a dynamic balance between growth inhibitory autocrine factors such as transforming growth factor β (TGFβ) and growth stimulating autocrine factors such as bFGF regulates vascular smooth muscle growth (15) and the evidence reviewed below supports this type of mechanism in thromboxane-stimulated vascular smooth muscle growth.

Roles of thromboxane in vascular smooth muscle cells

Thromboxane as a stimulus for vascular smooth muscle growth

As with other potent vasoconstrictors such as angiotensin II and thrombin (15,16), thromboxane stimulates vascular smooth muscle growth in tissue culture. Also as with other vasoconstricting growth promoters, the nature of the reported thromboxane-stimulated growth responses has varied with the species and particular vascular bed of cells origin and with the laboratory performing the work. In quiescent, low passage number cultured rat aortic smooth muscle cells, our laboratory finds that thromboxane stimulates hypertrophy, defined as an increase in protein synthesis (typically measured as incorporation of [^{3}H] leucine) and protein content (protein/DNA ratio) without significant DNA synthesis or increased cell number (17-19). This hypertrophic response is typical of that observed with other vasoconstricting growth promoters in truly quiescent vascular smooth muscle cell cultures and indicates that, compared to complete growth factors such as PDGF, thromboxane is an incomplete mitogen. The strong likelihood is that hypertrophic agonists such as thromboxane and angiotensin II activate a subset of cell signals which is sufficient for protein synthesis, but fail to activate other critical messages which are required for DNA synthesis and cell division. In a comparison of the relative ability of thromboxane (a hypertrophic agonist) and PDGF or the phorbol ester PMA (both hyperplastic agents) to activate individual components of the tyrosine kinase cascade, we found that only the hyperplastic agonists increased the phosphotyrosine content of GTPase activating protein, phospholipase C γ-1, and phosphatidylinositol 3-kinase (20). As these results are identical to those previously reported for angiotensin II (21), it appears that hypertrophy and hyperplasia can be determined, at least in part, by which proliferative cell signals are activated. In this regard it is likely that hypertrophy, rather than being a growth response separate and distinct from hyperplasia, is simply an incomplete portion of the proliferative cycle. Hypertrophy is hyperplasia which has progressed through the protein synthesis portion of the cell cycle, but which is aborted prior to DNA synthesis and cell division. This is consistent with observations by a variety of laboratories that thromboxane treatment is sufficient to increase expression of protooncogenes (early G1 phase) and to stimulate protein synthesis, but not DNA synthesis or cell proliferation (17-19,22).

It should be noted that in some cell preparations, such as guinea pig coronary artery-derived cells (23), and in studies of rat aortic cells from some laboratories (24,25), thromboxane appears to stimulate a hyperplastic or proliferative response. In fact, the initial description of thromboxane growth effects in vascular smooth muscle cells reported a mitogenic effect of the agonist U46619 in the presence of 1 μM insulin, but not in the absence of insulin (26). As insulin is a weak growth factor in this cell system, this report more correctly observed proliferative synergism between two relatively weak growth promoters. In any case, attempting to force a clear distinction between thromboxane-stimulated vascular smooth muscle cell hypertrophy and hyperplasia, while potentially useful for characterizing cellular responses in tissue culture, probably has less meaning *in vivo* where vascular smooth muscle cells are simultaneously exposed to multiple growth altering substances. This is especially true for reactive smooth muscle growth after vascular injury where a combination of platelet derived and endogenous growth factors, produced as a result of mechanical injury or release during cell lysis, undoubtedly act in concert to stimulate neointima formation. In this context it is interesting to note that thromboxane augments serum-stimulated, thrombin-stimulated, and platelet-derived growth factor-stimulated vascular smooth muscle growth in tissue culture (24,25,27). The mechanisms for this synergy have been examined in an interesting report by Nagata et al who analyzed the effects of the thromboxane agonist STA_2 on cell cycle progression in rat aortic vascular smooth muscle cells (24). Thromboxane increased DNA synthesis and shortened the doubling time of randomly cycling proliferating vascular smooth muscle cells. These effects were attributed to thromboxane-stimulated augmentation of the synthesis of actin and other cytoskeletal proteins during the S phase to G_2/M phase transition. Similar synergistic activity between thromboxane and thrombin has been noted in a recent preliminary report by Schrör and colleagues (27). Importantly, the potential for other platelet derived growth modulators, including serotonin and ADP, to act as amplification factors in platelet stimulated vascular smooth muscle cell growth is also great (28), and the growth promoting actions of multiple different platelet-derived substances has implications on the design of successful clinical approaches to interfering with platelet-stimulated vascular smooth muscle growth in the post-angioplasty setting.

Vascular smooth muscle thromboxane receptors

The exact nature of the human vascular smooth muscle thromboxane receptor has been, and continues to be, a matter of considerable controversy. There is compelling pharmacologic and functional evidence, in the form of differences in rank order agonist and antagonist potency, ligand binding affinity, agonist/antagonist activity, and patterns of heterologous desensitization, that different forms of thromboxane receptors are expressed in human vascular tissues compared to platelets (29-32). In this regard, the thromboxane agonist [125I]-BOP has been a particularly useful ligand for distinguishing between putative vascular and platelet thromboxane receptor sub-types as it binds to cultured human and rodent vascular smooth muscle cells with an affinity (approximately 400 pM) which is roughly an order of magnitude greater than its affinity for human platelet thromboxane receptors (32). However, molecular cloning of thromboxane receptors from multiple human tissues has thus far failed to identify a receptor exhibiting the pharmacologic characteristics of the putative human vascular sub-type. To date, complete or partial human thromboxane receptor cDNAs, designated TPα receptors, have been isolated from megakaryocytic cells (33,34), placenta (33), K562 chronic myelogenous leukemia cells (34), lung (35), and uterus (35), and each has [125I]-BOP binding characteristics of the human platelet type receptor which mediates the aggregation response (36). A splice variant of the human TPα receptor, differing only in the intracellular carboxyl terminus and designated TPβ, has been detected in human endothelial cells (37) and platelets (38) and has established the existence of distinct thromboxane receptor subtypes. However, the [125I]-BOP binding characteristics and the rank order binding affinities of several thromboxane analogs are identical in pure populations of TPα and TPβ receptors when transiently expressed in HEK293 cells (Dorn GW, unpublished observations). Thus, the existence and molecular characteristics of a unique human vascular thromboxane receptor remain uncertain at this time.

In contrast, the molecular nature of the rat vascular smooth muscle thromboxane receptor, the tissue and species utilized for most studies of thromboxane-stimulated vascular smooth muscle growth (18-20,24,26), has been elucidated. A rat thromboxane receptor was cloned from a kidney cDNA library (39) and this receptor is expressed in rat vascular smooth muscle cells as detected by *in situ* hybridization (39). Our laboratory has expressed the rat thromboxane receptor in HEK293 cells and finds that it exhibits high [^{125}I]-BOP binding affinity (40), thus reproducing the observations that a high affinity I-BOP receptor is expressed in cultured rat vascular smooth muscle cells (17). This binding pharmacology contrasts to the lower affinity I-BOP binding of the transfected human TPα and TPβ receptors (34, unpublished results). Thus, this is the first example of a cloned thromboxane receptor having the pharmacologic characteristics predicted for the putative vascular TP receptor subtype. Comparison of the predicted amino acid sequences of the human platelet TPα receptor and the rat thromboxane receptor reveal overall 74% identity and mostly highly conserved substitutions (Figure 1). Therefore, the unique pharmacology of the rat vascular receptor is the result of a relatively small number of structural determinants. Possession of two structurally similar receptors, each with unique ligand binding characteristics, will greatly facilitate experimental analysis of the structural determinants of thromboxane binding through engineering and expressing chimeric human/rat thromboxane receptors. Such studies have the potential to accelerate development of clinically useful thromboxane receptor subtype specific agonists/antagonists, and possibly to hasten the molecular cloning of the human vascular thromboxane receptor, if it exists.

Thromboxane receptor-stimulated growth signaling in vascular smooth muscle cells

While the molecular nature of human vascular smooth muscle thromboxane receptors is uncertain, much is known about thromboxane-stimulated cell signaling and functional responses in human and rodent vascular smooth muscle cells. As with other growth-promoting vasoconstrictors, thromboxane stimulates a pertussis toxin-insensitive activation of phospholipase C and release of intracellular free calcium (41), indicating receptor coupling to G-proteins of the

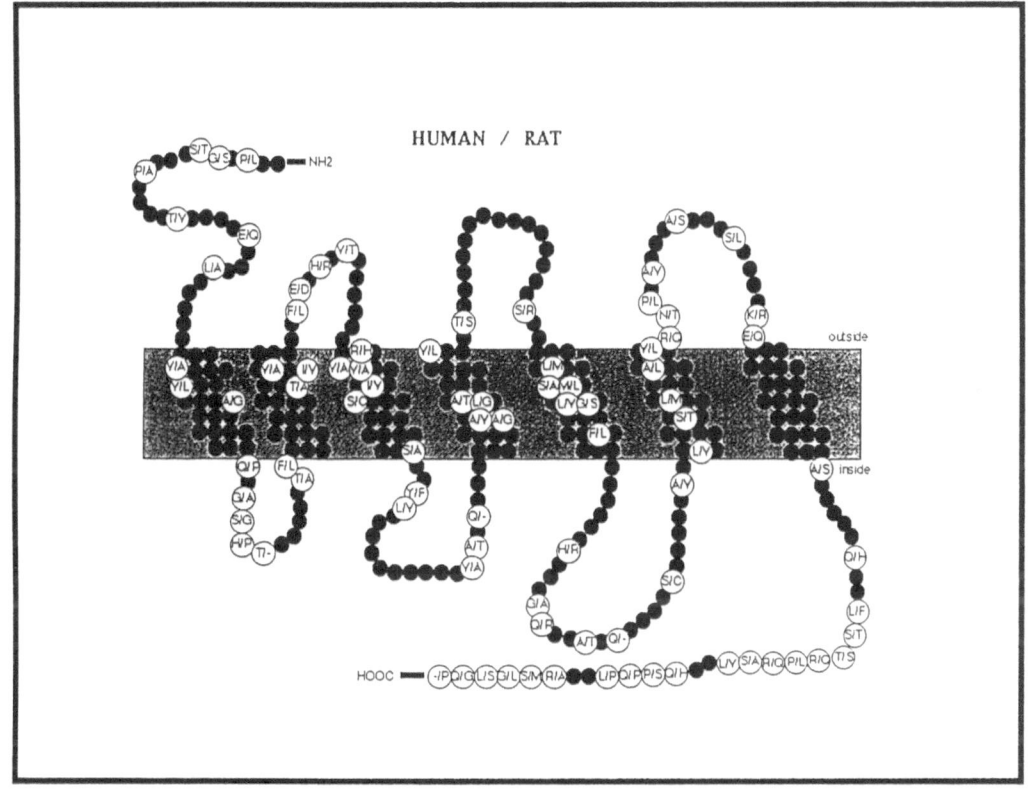

Figure 1. Comparison of amino acids in human TPα and rat TP receptor. Amino acids in the putative structure of the thromboxane receptor are depicted as closed circles if they are identical. The human/rat analogs are shown for non-identical amino acids. Amino acid identity is 74% overall and 100% in the seventh transmembrane domain.

Gq family. As noted above, thromboxane stimulation of quiescent vascular smooth muscle cells results in increased expression of *c fos*, *c myc*, and *egr-1* (17,25) protooncogenes which are associated with entry into the cell growth cycle. Thromboxane also causes tyrosine phosphorylation and activation of a number of cellular proteins (20), including MAP kinase (23). The exact mechanism(s) by which thromboxane stimulation of a Gq-coupled receptor activates the tyrosine kinase cascade have not yet been thoroughly examined. Three possible links between thromboxane receptors and tyrosine phosphorylation of cellular substrates exist and are depicted in Figure 2.

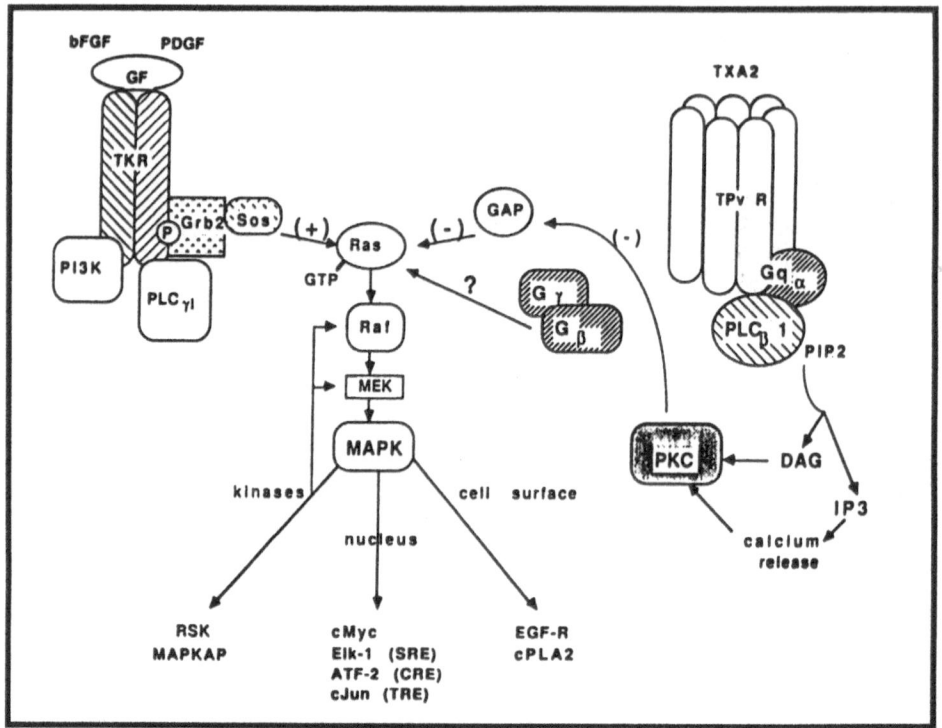

Figure 2: Simplified scheme for thromboxane-stimulated activation of MAP kinase. Thromboxane-stimulated protein kinase C (PKC) can inhibit GAP, thus disinhibiting ras and activating the tyrosine kinase cascade. Alternately, free $G_{\beta\gamma}$ units may directly activate tyrosine kinases. Finally, endogenous growth factors synthesized in response to thromboxane can initiate the cascade via their tyrosine kinase receptors.

First, thromboxane-stimulated activation of protein kinase C, via inhibition of ras-GTPase activating protein, can increase ras activity and activate the tyrosine kinase cascade. There is substantial evidence supporting activation of PKC as being necessary for thromboxane-stimulated vascular smooth muscle growth (19) and cross talk between PKC and tyrosine kinases exists at this level. A second possible mechanism is suggested by the observation that, like angiotensin II and thrombin, (15,16) thromboxane increases vascular smooth muscle synthesis of endogenous bFGF (18,19) which, if exported from the cell, could activate its own tyrosine kinase-coupled membrane receptor, thus initiating tyrosine phosphorylation events and stimulating cell growth. Inhibition of bFGF synthesis with anti-sense oligonucleotides prevents thromboxane-stimulated hypertrophy of serum deprived rat aortic smooth muscle cells (18). However, bFGF synthesis alone, in the absence of PKC activation, is not sufficient to cause thromboxane-stimulated growth (19). The third possible pathway coupling thromboxane receptors to tyrosine kinase reactions is for Gβ/γ subunits, released from activated Gq heterotrimers, to independently activate components of the tyrosine kinase cascade. Although there is evidence that Gβ/γ can activate the MAP kinase/tyrosine kinase cascade under certain circumstances (42,43), there is as yet no experimental support for the notion that Gβ/γ subunits affect thromboxane-mediated vascular smooth muscle growth. However, reagents are available to specifically test this hypothetical mechanism (43). A clearer picture of the potential relative contributions of thromboxane-stimulated tyrosine kinase activation, and the mechanisms for such activation, should be forthcoming.

Endogenous growth modifiers in thromboxane-stimulated vascular smooth muscle growth

As noted above, thromboxane stimulates vascular smooth muscle cell synthesis of endogenous bFGF, and this is a necessary, but not sufficient component of thromboxane-stimulated vascular smooth muscle growth. The evidence for the necessity of endogenously formed bFGF in thromboxane-stimulated growth is twofold. First, thromboxane stimulates increased endogenous expression of bFGF by approximately three hundred percent and preventing this reactive increase in bFGF through the application of antisense oligonucleotides directed at bFGF prevents thromboxane-stimulated hypertrophy (18). Second, application of an antibody which neutralizes the biological activity of bFGF to thromboxane-stimulated cells attenuates the resulting hypertrophy (18). The latter experiment not only supports the role of endogenous bFGF in thromboxane-stimulated growth, but indicates that at least some activity is the consequence of extracellular bFGF. Furthermore, the observation that anti-bFGF does not completely eliminate thromboxane-stimulated hypertrophy suggested that endogenous bFGF was also having intracellular effects. This latter notion has been confirmed in studies of cultured vascular smooth muscle cells derived from transgenic mice overexpressing human bFGF which were noted to be "hypertrophied" relative to cells from non-transgenic mice (44).

As bFGF is a true proliferative agonist when applied externally to vascular smooth muscle cells, and thromboxane causes bFGF formation and release, why isn't thromboxane-stimulated bFGF expression associated with hyperplasia rather than hypertrophy? A possible answer to this enigma was suggested by studies of angiotensin II-stimulated vascular smooth muscle hypertrophy, also associated with increased bFGF expression (15). Dzau and colleagues studied expression of transforming growth factor β_1 (TGFβ) and found that it was increased in angiotensin II stimulated vascular smooth muscle cells (15). As TGFβ is antiproliferative in this cell type, endogenously produced bFGF may have stimulated growth, but concomitant expression of TGFβ stopped the cell cycle short of DNA synthesis and mitosis. In preliminary studies we have examined this possibility in the context of thromboxane-stimulated vascular smooth muscle hypertrophy using smooth muscle cells cultured from the aortas of $TGF\beta_1$ knockout mice (45). In normal mouse vascular smooth muscle cells, as in rat cells, thromboxane caused cellular hypertrophy. However, in $TGF\beta_1$ knockout-derived cells, thromboxane caused hyperplasia. These studies provide additional support for the notion that vasoconstrictor-stimulated vascular smooth muscle growth in general is regulated by a dynamic balance of endogenously produced growth promoting and inhibiting factors. The relative roles of intracellular verses secreted forms of bFGF, and of different TGFβ types should be a fruitful area for future investigation.

Roles of other prostaglandins in vascular smooth muscle growth

It has been appreciated for some time that several arachidonic acid products have similar functional effects on their target tissues. For example, PGH_2 and thromboxane both bind to platelet thromboxane receptors with high affinity and have similar effects (46). Thus, specific inhibition of thromboxane synthesis, if accompanied by increased PGH_2, may not eliminate thromboxane stimulated events. Another cyclooxygenase product, prostaglandin $F_2\alpha$, has been widely employed as a vasoconstrictor, yet its vasoconstricting actions appear to be mediated almost exclusively via cross reactions at vascular thromboxane receptors. We and others have found a 20 fold discrepancy between the binding affinity of authentic PGF$_{2\alpha}$ and the extremely high concentrations necessary to achieve vascular contraction (17). Furthermore, vasoconstriction stimulated by PGF$_{2\alpha}$ is prevented by thromboxane receptor blockade whereas phospholipase C activation and increased intracellular calcium is not. These observations raise the question as to what, if not vasoconstriction, may be physiologically relevant functional consequence of PGF$_{2\alpha}$ on vascular smooth muscle? As PGF$_{2\alpha}$ interacts at its own receptor, activates PLC, increases intracellular free calcium, and stimulates vascular smooth muscle hypertrophy at nanomolar concentrations (17), it is possible that the primary function of this "vasoconstricting" prostaglandin is as a growth promoter rather than as a vasoconstrictor. This notion is supported by recent studies describing PGF$_{2\alpha}$-stimulated growth and growth signaling in neonatal rat ventricular myocytes (47) and in NIH-3T3 cells (48). Furthermore, as PGF$_{2\alpha}$ (but not thromboxane A$_2$) is produced by vascular endothelial cells (49), and is much more stable than thromboxane at physiologic temperature and pH, it has the potential to play an important role as a stimulus for, or synergistic factor in, vascular smooth muscle growth after chemical or mechanical endothelial injury.

Thromboxane receptor blockade and thromboxane synthesis inhibition as prophylaxis for post-angioplasty restenosis

It is clear from the preceding review of experimental studies that thromboxane at the very least can augment the vascular smooth muscle proliferative activity of peptide growth factors, and under some circumstances can act as an independent stimulus for vascular smooth muscle hypertrophy or proliferation. Since thromboxane is released in large amounts from activated platelets, and platelet activation is inherent in percutaneous revascularization procedures which necessarily consist of intimal and medial disruption, exposure of collagen to circulating platelets, and local release of multiple platelet activating factors, it is reasonable to conclude that interfering with the growth promoting effects of thromboxane could have beneficial effects on angioplasty restenosis. This formed the rationale for clinical trials of pharmacologic inhibitors of thromboxane synthesis or receptor blockers after coronary angioplasty.

Thromboxane synthesis inhibition can be achieved with several agents suitable for therapeutic use. A hypothetical advantage of selective inhibition of thromboxane synthesis, rather than inhibition of the entire cyclooygenase cascade with such agents as aspirin, is that endoperoxides no longer metabolized to thromboxane would be redirected in the metabolic pathway toward synthesis of beneficial prostaglandins such as prostacyclin (see Reference 50 for review). However a clinical trial of the thromboxane synthesis inhibitor CV-4151 failed to detect any change in the incidence of restenosis three to six months post angioplasty (51). There are at least two reasons why this result might have been expected. First, the immediate precursor to thromboxane which would accumulate after thromboxane synthesis inhibition is the endoperoxide PGH_2 which, as noted above, interacts at thromboxane receptors with an affinity at least equal to thromboxane A_2 (46) and which appears to have functional activity identical to thromboxane. In fact, most of the studies performed in tissue culture which report "thromboxane-stimulated" vascular smooth muscle growth have employed the stable endoperoxide analog U46619 as a stimulus rather than utilizing a true structural analog of thromboxane. The second possible reason for failure of thromboxane synthesis inhibition to prevent angioplasty restenosis is that endothelial denudation after balloon dilatation removes the cell type which would be primarily responsible for synthesis of antimitotic prostaglandins using the "endoperoxide shunt".

An alternate approach is the use of thromboxane receptor blockade to prevent the target cell effects of both thromboxane and prostaglandin endoperoxides on vascular smooth muscle cells. Two large clinical trials have tested thromboxane receptor antagonists after angioplasty. The first is the CARPORT study, published in 1991 (52). The thromboxane antagonist GR32191 or a placebo was administered to patients before and for six months after angioplasty and the incidence of restenosis was analyzed at follow-up by quantitative coronary angiography. In this study there was no difference in coronary diameter, myocardial infarction, or the need for surgical or repeat percutaneous revascularization between treated and control groups. Of note, patients in the CARPORT study did not take aspirin.

In contrast, the recently completed M-HEART II trial compared thromboxane receptor blockade with sulotroban to non-selective thromboxane synthesis inhibition using aspirin (53). Compared to placebo treated controls, both aspirin and sulotroban decreased the incidence of myocardial infarction post angioplasty. However, neither agent affected the incidence of restenosis compared to placebo and, when compared to aspirin treatment, sulotroban treatment may actually have increased the incidence of restenosis (53% versus 39%). Overall clinical outcome was statistically better in the aspirin treated group than in the group treated with the thromboxane receptor blocker.

A clinical role for thromboxane in vascular disease?

With newer approaches to angioplasty restenosis, including intracoronary stents and the broad spectrum antiplatelet agents of the anti GpIIb/IIIa category having significant impacts on restenosis, current practice should be associated with clinical restenosis rates of 10 to 20 % (7,54,55). The future use of intracoronary stents as both mechanical vascular support devices and as vehicles for the delivery of antiproliferative therapies such as heparin (56) or beta radiation (57,58) will undoubtedly further lower restenosis rates. These therapies clearly are succeeding where targeting thromboxane or other individual growth promoting factors has not. The success of non-specific treatments such as mechanical stenting and anti GpIIb/IIIa agents has helped to further our understanding about restenosis pathophysiology. Namely, restenosis is the end result of vascular growth resulting from multiple divergent chemical and mechanical stimuli. Platelets themselves release peptide growth factors, thromboxane, serotonin, and ADP, and each of these agents can stimulate vascular smooth muscle cell growth. Mechanical stretch of vascular smooth muscle cells is itself a stimulus for growth, independent of exogenous factors (59-61). Endothelial injury not only eliminates the potentially beneficial actions of prostacyclin, but can increase the synthesis and release of proliferative prostaglandins such as PGF$_{2\alpha}$ (49). Finally, circulating thrombin clearly has mitotic effects on vascular smooth muscle cells (16). Therefore, it is likely that therapy targeted at thromboxane's growth promoting effects will ultimately be only a single facet of a successful multiple approach strategy for inhibiting neointimal proliferation after balloon dilatation or atherectomy. Thromboxane receptor blockade, synthesis inhibition, or combined therapy may prove to have more significant impact on vascular abnormalities in other diseases such as primary pulmonary hypertension.

Acknowledgements
Supported by: Grants HL49267 and P50-HL52318-01 from The National Institutes of Health, and a Merit Review grant from the Veterans Administration. Gerald W. Dorn II, M.D. is an Established Investigator of the American Heart Association (supported with funds contributed in part by its Ohio Affiliate).

References

1.   Gordon D, Reidy MA, Benditt EP, Schwartz SM. Cell proliferation in human coronary arteries. Proc Natl Acad Sci USA 1990; 87:4600-4604.

2.   Johnson DE, Hinohara T, Selmon MR, Braden LJ, Simpson JB. Primary peripheral arterial stenoses and restenoses excised by transluminal atherectomy: A histopathologic study. J Am Coll Cardiol 1990; 425:419-425.

3.   Alkjaer C, Heagerty AM, Peterson KK, Swales JD, Mulvany MJ. Evidence for increased media thickness, increased neuronal amine uptake, and depressed excitation-contraction coupling in isolated resistance vessels from essential hypertensives. Circ Res 1987; 61:181-186.

4.   Haworth SG. Primary pulmonary hypertension. Br Heart J 1983; 49:517-521.

5.   Johnson DE, Alderman EL, Schroeder JS, Gao S-Z, Hunt S, DeCampli WM, Stinson E, Billingham M. Transplant coronary artery disease: Histopathologic correlations with angiographic morphology. J Am Coll Cardiol 1991; 17:449-457.

6.   Fingerle J, Johnson R, Clowes AW, Majesky MW, Reidy MA. Role of platelets in smooth muscle cell proliferation and migration after vascular injury in rat carotid artery. Proc Natl Acad Sci USA 1989; 86:8412-8416.

7.   Topol EJ, Califf RM, Weisman HF, Ellis SG, Tcheng JE, Worley S, Ivanhoe R, George BS, Fintel D, Weston M. et al. Randomized trial of coronary intervention with antibody against platelet IIb/IIIa integrin for reduction of clinical restenosis: results at six months. Lancet 1994; 343:881-886.

8. Nagai R, Larson DT, Periasamy M. Characterization of a mammalian smooth muscle myosin heavy chain cDNA clone and its expression in various smooth muscle types. Proc Natl Acad Sci USA 1988; 85:1047-1051.

9. Owens G, Schwartz S. Alterations in vascular smooth muscle mass in the spontaneously hypertensive rat: Role of cellular hypertrophy, hyperploidy, and hyperplasia. Circ Res 1982; 51:280-289.

10. Schwartz S, Campbell GR, Campbell JH. Replication of smooth muscle cells in vascular disease. Circ Res 1986; 58:427-444.

11. Owens G, Reidy M. Hyperplastic growth response of vascular smooth muscle cells following induction of acute hypertension in rat by aortic coarctation. Circ Res 1985; 57:695-705.

12. Lindner V, Reidy MA. Proliferation of smooth muscle cells after vascular injury is inhibited by an antibody against basic fibroblast growth factor. Proc Natl Acad Sci USA 1991; 88:3739-3743.

13. Ferns GAA, Raines EW, Sprugel KH, Motani AS, Reidy MA, Ross R. Inhibition of neointimal smooth muscle accumulation after angioplasty by an antibody to PDGF. Science 1991; 253:1129-1133.

14. Galloway AC. Suppression of neointimal lesions after vascular injury: A role for polyclonal anti-basic fibroblast growth factor antibody. Surgery 1994; 116:456-462.

15. Itoh H, Mukoyama M, Pratt RE, Gibbons GH, Dzau VJ. Multiple autocrine growth factors modulate vascular smooth muscle cell growth response to angiotensin II. J Clin Invest 1993; 91:2268-2274.

16. Weiss RH, Maduri M. The mitogenic effect of thrombin in vascular smooth muscle cells is largely due to basic fibroblast growth factor. J Biol Chem 1993; 268:5724-5727.

17. Dorn GW II, Becker MW, Davis MG. Dissociation of the contractile and hypertrophic effects of vasoconstrictor prostanoids in vascular smooth muscle. J Biol Chem 1992; 267:24897-24905.

18. Ali S, Davis MG, Becker MW, Dorn GW II. Thromboxane A$_2$ stimulates vascular smooth muscle hypertrophy by upregulating the synthesis and release of endogenous basic fibroblast growth factor. J Biol Chem 1993; 268:17397-17403.

19.   Ali S, Becker MW, Davis MG, Dorn GW II. Dissociation of vasoconstrictor-stimulated basic fibroblast growth factor expression from hypertrophic growth in cultured vascular smooth muscle cells: Relevant roles of protein kinase C. Circ Res 1994; 75:836-843.

20.   Ali S, Dorn GW II. Patterns of tyrosine phosphorylation differ in vascular hypertrophy and hyperplasia. Am J Physiol 1994; 36:C1674-C1681.

21.   Molloy CJ, Taylor DS, Weber H. Angiotensin II stimulation of rapid protein tyrosine phosphorylation and protein kinase activation in rat aortic smooth muscle cells. J Biol Chem 1993; 268:7338-7345.

22.   Zucker TP, Grosser T, Morinelli T, Halushka PV, Sachinidis A, Vetter H, Schrör K. Potentiation of PDGF-induced growth responses in coronary artery smooth muscle cells by thromboxane. Agents Actions 1995; 45:53-58.

23.   Morinelli TA, Zhang L-M, Newman WH, Meier KE. Thromboxane A_2/prostaglandin H_2-stimulated mitogenesis of coronary artery smooth muscle cells involves activation of mitogen-activated protein kinase and S6 kinase. J Biol Chem 1994; 269:5693-5698.

24.   Nagata T, Uehara Y, Numabe A, Ishimitsu T, Hirawa N, Ikeda T, Matsuoka H, Sugimoto T. Regulatory effect of thromboxane A_2 on proliferation on vascular smooth muscle cells from rats. Am J Physiol 1992; 263:H1331-H1338.

25.   Sachinidis A, Flesch M, Ko Y, Schrör K, Bohm M, Düsing R, Vetter H. Thromboxane A_2 and vascular smooth muscle cell proliferation. Hypertension 1995; 26:771-780.

26.   Hanasaki K, Nakano T, Arita H. Receptor-mediated mitogenic effect of thromboxane A_2 in vascular smooth muscle cells. Biochem Pharmacol 1990; 40:2535-2542.

27.   Zucker T-P, Bönisch D, Schrör K. Thromboxane potentiates thrombin- and thrombin receptor activating peptide (TRAP)-induced mitogenesis of coronary artery smooth muscle cells. Circulation 1995; 92:0522.

28.   Crowley ST, Dempsey ED, Horwitz KB, Horwitz LD. Platelet-induced vascular smooth muscle cell proliferation is modulated by the growth amplification factors serotonin and adenosine diphosphate. Circulation 1994; 90:1908-1918.

29.   Mais DE, Saussy DL KR, Chaikhouni A, Kochel PJ, Knapp DR, Hamanaka N, Halushka PV. Pharmacologic characterization of human and canine thromboxane A_2/prostaglandin H_2 receptors in platelets and blood vessels: Evidence for different receptors. J Pharmacol Exp Ther 1985; 233:418-424.

30.   Morinelli TA, Okwu AK, Mais DE, Halushka PV, John V, Chen C-K, Fried J. Difluorothromboxane A$_2$ and stereoisomers: Stable derivatives of thromboxane A$_2$ with differential effects on platelets and blood vessels. Proc Natl Acad Sci USA 1989; 86:5600-5604.

31.   Furci L, Fitzgerald DJ, Fitzgerald GA. Heterogeneity of prostaglandin H$_2$/thromboxane A$_2$ receptors: Distinct subtypes mediate vascular smooth muscle contraction and platelet aggregation. J Pharmcol Exp Ther 1991; 258:74-81.

32.   Dorn GW II. Tissue- and species-specific differences in ligand binding to thromboxane A$_2$ receptors. Am J Physiol 1991; 261:R145-R153.

33.   Hirata M, Hayashi Y, Ushikubi F, Yokota Y, Kageyama R, Nakanishi S, Narumiya S. Cloning and expression of cDNA for a human thromboxane A$_2$ receptor. Nature 1991; 349:617-620.

34.   D'Angelo DD, Davis MG, Ali S, Dorn GW II. Cloning and pharmacologic characterization of a thromboxane A$_2$ receptor from K562 (Human Chronic Myelogenous Leukemia) Cells. J Pharmacol Exp Ther 1994; 271:1034-1041.

35.   D'Angelo DD, Davis MG, Houser WA, Eubank JJ, Ritchie ME, Dorn GW II. Characterization of the 5'end of the human thromboxane receptor gene: Organizational analysis and mapping of protein kinase C responsive elements regulating expression in platelets. Circ Res 1995; 77:466-474.

36.   Dorn GW II. Distinct platelet thromboxane A$_2$/prostaglandin H$_2$ receptor subtypes. A radioligand binding study of human platelets. J Clin Invest 1989; 84:1883-1891.

37.   Raychowdhury MK, Yukawa M, Collins LJ, McGrail SH, Kent KC, Ware JA. Alternative splicing produces a divergent cytoplasmic tail in the human endothelial thromboxane A$_2$ receptor. J Biol Chem 1994; 269:19256-19261.

38.   Hirata T, Ushikubi F, Kakizuka A, Okuma M, Narumiya S. Two thromboxane A$_2$ receptor isoforms in human platelets. Opposite coupling to adenylyl cyclase with different sensitivity to Arg60 to Leu mutation. J Clin Invest 1996; 97:949-956.

39.   Abe T, Takeuchi K, Takahashi N, Tsutsumi E, Taniyama Y, Abe K. Rat kidney thromboxane receptor: Molecular cloning, signal transduction, and intrarenal expression localization. J Clin Invest 1995; 96:657-664.

40.   D'Angelo DD, Terasawa T, Carlisle SJ, Dorn GW II, Lynch KR. Characterization of a rat kidney thromboxane A$_2$ receptor: High affinity for the agonist ligand I-BOP. Prostaglandins 1996; 52:303-316.

41.   Dorn GW II, Becker MW. Thromboxane A_2 stimulated signal transduction in vascular smooth muscle. J Pharmacol Exp Ther 1993; 265:447-456.

42.   van Biesen T, Hawes BE, Luttrell DK, Krueger KM, Touhara K, Porfiri E, Sakaue M, Luttrell LM, Lefkowitz RJ. Receptor-tyrosine-kinase-and $G\beta/\gamma$-mediated MAP kinase activation by a common signaling pathway. Nature 1995; 376:781-784.

43.   Zhang J, Zhang J, Shattil SJ, Cunningham MC, Rittenhouse SE. Phosphoinositide 3-kinase γ and p85/phosphoinositide 3-kinase in platelets. J Biol Chem 1996; 271:6265-6272.

44.   Coffin JD, Florkiewicz RF, Neuman J, Hopkins TM, Dorn GW II, Lightfoot P, German R, Howles PN, Kier A, O'Toole BA, Sasse J, Gonzalez AM, Baird A, Doetschman TC. Abnormal bone growth and selective translational regulation in basic fibroblast growth factor (FGF-2) transgenic mice. J Cell Biol 1995; 6:1861-1873.

45.   Ali S, Doetschman T, Dorn GW II. Transforming growth factor β_1 knockout converts thromboxane stimulated vascular hypertrophy to hyperplasia. Circulation 1994; 90:749.

46.   Mayeux PR, Morton HE, Gillard J, Lord A, Morinelli TA, Boehm A, Mais DE, Halushka PV. The affinities of prostaglandin H_2 and thromboxane A_2 for their receptor are similar in washed human platelets. Biochem Biophys Res Commun 1988; 157:733-739.

47.   Adams JW, Migita DS, Yu MK, Young R, Hellickson MS, Castro-Vargas FE, Domingo JD, Lee PH, Bui JS, Henderson SA. Prostaglandin $F_{2\alpha}$ stimulates hypertrophic growth of cultured neonatal rat ventricular myocytes. J Biol Chem 1996; 271:1179-1186.

48.   Watanabe T, Nakao A, Emerling D, Hashimoto Y, Tsukamoto K, Horie Y, Kinoshita M, Kurokawa K. Prostaglandin $F_{2\alpha}$ enhances tyrosine phosphorylation and DNA synthesis through phospholipase C-coupled receptor via Ca^{2+}-dependent intracellular pathway in NIH-3T3 cells. J Biol Chem 1994; 269:17619-17625.

49.   Revtyak GE, Johnson AR, Campbell WB. Prostaglandin synthesis in bovine coronary endothelial cells: Comparison with other commonly studied endothelial cells. Thromb Res 1987; 48:671-683.

50.   Gresele P, Deckmyn H, Nenci GG, Vermylen J. Thromboxane synthase inhibitors, thromboxane receptor antagonists and dual blockers in thrombotic disorders. TiPS 1991; 12:158-163.

51.   Hattori R, Kodama K, Takatsu F, Yui Y, Kawai C. Randomized trial of a selective inhibitor of thromboxane A_2 synthetase, (E)-7-phenyl-7-(3-pyridyl)-6-heptenoic acid (CV-4151), for prevention of restenosis after coronary angioplasty. Jap Circ J 1991; 55:324-329.

52. Serruys PW, Rutsch W, Heyndrickx GY, Danchin N, Mast EG, Wijns W, Rensing BJ, Vos J, Stibbe J. Prevention of restenosis after percutaneous transluminal coronary angioplasty with thromboxane A$_2$-receptor blockade. A randomized, double-blind, placebo-controlled trial. Circulation 1991; 84:1568-1580.

53. Savage MP, Goldberg S, Bove AA, Deutsch E, Vetrovec G, Macdonald RG, Bass T, Margolis JR, Whitworth HB, Taussig A, Hirshfeld JW, Cowley M, Hill JA, Marks RG, Fischman DL, Handberg E, Herrmann H, Pepine CJ. Effect of thromboxane A$_2$ blockade on clinical outcome and restenosis after successful coronary angioplasty. Multi-hospital eastern atlantic restenosis trial (M-Heart II). Circulation 1995; 92:3194-3200.

54. Serruys PW, de Jaegere P, Kiemeneij F, Macaya C, Rutsch W, Heyndrickx G, Emanuelsson H, Marco J, Legrand V, Materne P, et al. A comparison of balloon-expandable-stent implantation with balloon angioplasty in patients with coronary artery disease. N Eng J Med 1994; 331:489-495.

55. Macaya C, Serruys PW, Ruygrok P, Suryapranata H, Mast G, Klugmann S, Urban P, den Heijer P, Koch K, Simon R, Morice MC, Crean P, Bonnier H, Wijns W, Danchin N, Bourdonnec C, Morel MA. Continued benefit of coronary stenting versus balloon angioplasty: one-year clinical follow-up of Benestent trial. JACC 1996; 27:255-261.

56. Serruys PW, Emanuelsson H, van der Giessen W, Lunn AC, Kiemeney F, Macaya C, Rutsch W, Heyndrickx G, Suryapranata H, Legrand V, Goy JJ, Materne P, Bonnier H, Morice MC, Fajadet J, Belardi J, Colombo A, Garcia E, Ruygrok P, de Jaegere P, Morel MA. Heparin-coated Palmaz-Schatz stents in human coronary arteries. Early outcome of the Benestent-II Pilot Study. Circulation 1996; 93:412-422.

57. Hehrlein C, Gollan C, Donges K, Metz J, Riessen R, Fehsenfeld P, von Hodenberg E, Kubler W. Low-dose radioactive endovascular stents prevent smooth muscle cell proliferation and neointimal hyperplasia in rabbits. Circulation 1995; 92:1570-1575.

58. Hehrlein C, Stintz M, Kinscherf R, Schlosser K, Huttel E, Friedrich L, Fehsenfeld P, Kubler W. Pure beta-particle-emitting stents inhibit neointima formation in rabbits. Circulation 996; 93:641-645.

59. Sudhir K, Wilson E, Chatterjee K, Ives HE. Mechanical strain and collagen potentiate mitogenic activity of angiotensin II in rat vascular smooth muscle cells. J Clin Invest 1993; 92:3003-3007.

60. Davis MG, Ali S, Leikauf GD, Dorn GW II. Tyrosine kinase inhibition prevents deformation-stimulated vascular smooth muscle growth. Hypertension 1994; 24:706-713.

61.   Wilson E, Mai Q, Sudhir K, Weiss RH, Ives HE. Mechanical strain induces growth of vascular smooth muscle cells via autocrine action of PDGF. J Cell Biol 1993; 123:741-747.

AAS 48
Prostaglandins and Control of Vascular
Smooth Muscle Cell Proliferation
© 1997 Birkhäuser Verlag Basel

Roles of vasodilatory prostaglandins in mitogenesis of vascular smooth muscle cells

Karsten Schrör and Artur-Aron Weber

Institut für Pharmakologie, Heinrich-Heine-Universität Düsseldorf, Moorenstr. 5, D-40225 Düsseldorf, Germany
Phone: (+49-211) 81 12500; Fax: (+49-211) 81 14781; E-Mail: schroer@pharma.uni-duesseldorf.de

Summary. Vasodilatory prostaglandins (PGI_2, PGE_1) and synthetic prostacyclin mimetics inhibit smooth muscle cell proliferation in vitro after stimulation by growth factors. Similar results are obtained in vivo after endothelial injury, suggesting that vasodilatory prostaglandins might also control smooth muscle cell proliferation in vivo. However, available data from clinical trials are conflicting and currently do not support the concept that these compounds might be successfully used to suppress excessive smooth muscle cell growth in response to tissue injury, specifically restenosis after PTCA. One possible explanation for these different results is an agonist-induced down-regulation of prostacyclin receptors in vascular smooth muscle cells. It is possible that enhanced endogenous prostacyclin biosynthesis, subsequent to induction of COX-2 and/or in relation to the formation of a neointima from media smooth muscle cells, might have a similar effect. There is still uncertainty regarding the cellular signal transduction pathways and their possibly complex interaction, although cAMP-dependent reactions are probably involved. In addition, vasodilatory prostaglandins might also interfere with the generation and action of other growth modulating factors, including PDGF, hepatocyte growth factor and nitric oxide. In conclusion, vasodilatory prostaglandins might be considered as growth modulating endogenous mediators in vascular smooth muscle cells.

Introduction

Smooth muscle cell proliferation as a component of tissue repair in wound healing and atherosclerotic wall thickening

Subsequent to an injury to the endothelial lining of an artery, smooth muscle cells from the media migrate to the intima, proliferate and form a neointima (1). Processes similar to traumatic vessel injury also occur in atherosclerotic lesion formation (2). Uncontrolled smooth muscle cell growth and proliferation will cause vessel narrowing, eventually resulting in critical reduction of the vessel lumen and functionally relevant stenosis. These alterations will not occur in the presence of a functioning endothelium, suggesting that endothelium-derived factors are involved in the control of cell growth and division. Thus, a better understanding of the mechanisms that control smooth muscle cells function is not only of scientific interest in vascular biology but also has a significant clinical impact because so far no drugs are available that would prevent uncontrolled smooth muscle cells growth responses subsequent to balloon angioplasty or vascular stenting in patients suffering from vascular diseases (3).

One unique property of vascular smooth muscle cells is their plasticity, including their replicative potential. Specifically, smooth muscle cells are capable of expressing different phenotypes in vivo (1, 4). This phenotypic modulation, subsequent to injury, is a transformation from the differentiated „contractile" cell that responds to vasoactive factors and serves to control local perfusion into a less differentiated „synthetic" cell type which looses contractile filaments, produces extracellular matrix and reacts to mitogens with cell division (4-5). Similar phenotypic alterations of smooth muscle cells occur in tissue culture in the presence of growth factors and can be prevented by coculturing smooth muscle cells with endothelial cells (1, 4). This suggests that not only the contractile state of smooth muscle cells, i.e. the control of local perfusion, is under control of the endothelium but also the long-term adjustment of this regulation by control of mitogenesis and smooth muscle cell phenotype.

Growth, i.e. hypertrophy and/or hyperplasia of smooth muscle cells in response to injury is regulated by humoral and cellular factors. Specifically, vascular cells themselves are able to regulate their cytoskeletal structures and differentiation state by generating positive and negative differentiation signals. There is evidence suggesting that vasoconstrictor compounds, such as angiotensin II, endothelin or thromboxane A_2 act as growth promoting agents while vasodilator compounds, such as isoproterenol, nitric oxide or prostacyclin have the opposite effect (6-7). However, little is known regarding the generation and action of these mediators and the complex interplay between these autacoids (8-9).

Sjolund and colleagues (10) have described a phenotype modulatory action of PGE_1 in primary cultures of arterial smooth muscle cells. Larrue and colleagues (11) have shown that generation of PGI_2 by vascular smooth muscle cells shows a time-dependent increase over weeks after endothelial denudation and that this process is paralleled by the generation of a neointima from smooth muscle cells. These and other data have led to the conclusion that endogenous prostaglandins might serve different purposes in vascular smooth muscle cells: Maintenance of the differentiated, contractile state and control of proliferation in the presence of growth factors (12). This paper reviews the role of vasodilatory prostaglandins in control of smooth muscle cell mitogenesis. Three major issues are addressed: (i) Action of vasodilatory prostaglandins on smooth muscle cell proliferation; (ii) sources of vasodilatory prostaglandins, specifically vascular generation of these autacoids and (iii) intracellular signal transduction mechanisms. The hypothesis is put forward that the actions of vasodilatory prostaglandins, i.e. PGI_2 and its analogues and PGE_1, are bimodal in nature (i) stimulation of vascular smooth muscle cells to enter the cell cycle and to divide in response to tissue injury in order to facilitate tissue repair but also (ii) to prevent uncontrolled growth of vascular smooth muscle cells by interfering with the synthesis phase of the cell cycle.

Actions of PGI₂, PGE₁ and PGI₂ mimetics on proliferation of vascular smooth muscle cells

Cell culture

The first direct evidence for a growth inhibitory action of prostaglandins in vascular smooth muscle cells was published twenty years ago (13). Since then, the control of smooth muscle cells proliferation by vasodilatory prostaglandins, mainly PGE_1 and PGI_2, was studied extensively in vitro in vascular smooth muscle cells of human and animal origin (14-22). The dominating finding was inhibition of cell proliferation. Synthetic, chemically more stable PGI_2 mimetics, such as carbacylin and carbacyclin analogues, iloprost, cicaprost, ciprostene, beraprost and others were also found to inhibit proliferation of vascular smooth muscle cells after stimulation with growth factors (21, 23-26).

There are only few reports that could not confirm an inhibitory action of vasodilatory prostaglandins on growth responses in cultured smooth muscle cells. For example, Pasricha et al (27) did not observe any growth inhibitory effect of PGE_1 in serum-stimulated cultured bovine aortic smooth muscle cells under conditions, when the mitogenesis of pulmonary arterial smooth muscle cells was significantly enhanced. Stimulation of smooth muscle cells proliferation rather than inhibition by a variety of prostaglandins, including PGE_2 and PGI_2, was also reported by Palmberg et al (28). Thus, with a few exceptions, the general impression is that vasodilatory prostaglandins act as antiproliferative signals and antagonize the mitogenic action of growth factors on vascular smooth muscle cells.

Generalizations from cell culture studies have been done with much care since species variations, different experimental protocols, specifically the presence or absence of (defined) growth factors to stimulate DNA synthesis will markedly influence growth responses (4). The use of passaged cells instead of primary cultures might change activities of important enzymes (29), the degree of (de)differentiation modifies the responsiveness of important second messenger systems, such as cAMP (see below), to eicosanoids, in particular PGI_2 (30). It has also to be considered that smooth muscle cells proliferating under the influence of growth factors generally are in the secretory phenotype (4).

Animal studies

Antiproliferative actions of vasodilatory prostaglandins were also observed in animal experiments in vivo. Oral treatment of cholesterol-fed rabbits with cicaprost at a non-hypotensive dose for 3 months was found to significantly reduce aortic plaque formation and intima thickening as well as to improve endothelium-dependent vasodilation in both aortic tissue and coronary arteries (31). Thus, treatment with a PGI_2 mimetic, did not only modify smooth muscle cell proliferation but also had beneficial effects on the endothelium. TFC-132, another synthetic PGI_2 analogue, was also found to reduce neointimal thickening after oral administration to rabbits (26). Beraprost, given subcutaneously to cholesterol-fed rabbits, inhibited restenosis after balloon injury of the femoral artery (32). It is interesting to note that these beneficial effects in cholesterol-fed animals occurred in the absence of any reduction of the markedly (about 10-20-fold) elevated plasma cholesterol levels. This agrees with data of Orekhov et al (33) who demonstrated a significant inhibition of [^{3}H]-thymidine uptake, i.e. DNA synthesis, in cultured intimal smooth muscle cells from the human aorta. Interestingly, Levitt et al (34) did not detect any reduction of intimal hyperplasia in a rat model of carotid endothelial injury after administration of iloprost. In this study, rats were treated through an observation period of 2-weeks with continuous i.v. infusion of the prostacyclin mimetic. It is not clear whether the dose used, i.e. 100 ng/kg x min, was hypotensive in these animals, eventually resulting in an activation of sympathoadrenal pressor systems. There was also no prevention of restenosis subsequent to thermal laser arterial injury in dogs at 8 weeks after short-term intraarterial PGI_2 at the time of the surgical intervention. This therapeutic failure occurred despite of a more than 70% reduction in initial platelet deposition at the site of injury (35). Recently, gene transfer of the COX-1 gene was reported to enhance PGI_2 synthesis 4-fold and to prevent arterial thrombosis after carotid angioplasty in a pig model of carotid angioplasty (36). Whether this approach also results in prevention of restenosis, remains to be shown.

Taken together, several animal studies tend to confirm cell culture and in vitro data, suggesting that exogenous administration of prostacyclins inhibits intimal tickening and atherosclerotic plaque formation. However, negative results have been reported mainly for short-term administration and there was no clear connection between prevention of platelet adhesion and antimitogenic actions of prostaglandins. These actions of prostacyclins are not restricted to vascular smooth muscle cells but also involve improved preservation of endothelial function and can be separated from prostaglandin-related effects on tissue cholesterol levels and vascular smooth muscle relaxation, i.e. hypotensive actions (31).

Clinical trials

More detailed information on the action of vasodilatory prostaglandins on smooth muscle cell proliferation in clinical trials is the subject of another chapter of this book and is here only covered shortly with particular emphasis to the specificity of antimitogenic actions. Sinzinger and colleagues (19) have measured the number of activated smooth muscle cells in the plaque and media of femoral and popliteal arteries of patients prior to surgery (amputation) because of end-stage peripheral arterial occlusive disease. The percentage of activated smooth muscle cells, as determined from their microscopic appearance, tended to be reduced in patients who were treated with PGI_2 infusions for 6 h during 5 days prior to surgery. In patients with carotid atheromatous lesions, oral administration of the PGI_2 mimetic TRK-100 for 4 weeks caused a significant inhibition of platelet accumulation in carotid atheromas but no significant reduction in plaque size (38). In a prospective, double-blind trial, Darius and colleagues (39) found an inhibition of restenosis at 6 months after administration of ciprostene before and during the first 4 days after PTCA. In contrast, no inhibition of restenosis was seen in patients treated with intravenous infusions of PGI_2 for 3-4 days subsequent to PTCA (40-41). The interesting point, taken from these studies for further discussion, is the possible separation of antiplatelet and antimitogenic actions of prostaglandins.

Vascular prostaglandin production in tissue injury and atherosclerosis

Regulation of prostaglandin formation by vascular smooth muscle cells

Proliferating vascular cells synthesize more PGI_2 than quiescent cells (11, 30, 42) and smooth muscle cells appear to be the dominant source of PGI_2 production because of their large content of PGI-synthase (43). There are two major sources for prostaglandin production by vascular cells: the endothelium and smooth muscle cells. The major prostaglandin synthesized from both cell types is prostacyclin (20, 44). The amount of prostaglandins synthesized by vascular cells that are exposed to growth factors, such as PDGF or thrombin, is determined by the activity of the inducible isoform of the cyclooxygenase, COX-2 (45). Several cis-regulatory elements have been identified within the COX-2 promoter (7, 46-47). In proliferating cultured rat vascular smooth muscle cells, there was an about 10-fold higher PGI_2 formation as compared to quiescent cells. A large peak of this PGI_2 synthesis was seen at the G_o/G_1 transition state and was coincident with a peak in phospholipase A_2 activity which was preceded by an increase of phosphoinositide turnover and phospholipase A_2 mRNA (48). Additionally, up-regulation of vascular LDL-receptors after stimulation with PDGF will provide additional free arachidonic acid for PGI_2 synthesis (49). Interestingly, cytokines, such as interleukin-1 (50) and growth factors (51) not only induce proliferation of vascular smooth muscle cells, but also require endogenous prostaglandins for this effect. Their mitogenic actions are largely inhibited after blockade of endogenous prostaglandin formation. Interestingly, stimulation of endogenous prostacyclin biosynthesis has also been shown to reduce intimal thickening and plaque formation in cholesterol-fed rabbits (37).

Animal studies in vivo

Tissue injury is also associated with significant alterations in vascular prostaglandin biosynthesis. Overexpression of prostacyclin synthase in cultured neointimal smooth muscle cells resulted in a doubling of PGI_2 production and significant inhibition of DNA synthesis subsequent to stimulation with serum (52). Endothelial injury, for example in early atherogenesis (53, 54) or after ballooning of a large artery (55-58), is associated with an initital reduction of prostacyclin generation ex vivo. The initial fall is probably due to the removal of endothelium with its high COX protein content (60). However, migration of smooth muscle cells into the intima, i.e. formation of a neointima within some weeks, results in the production of increasing amounts of PGI$_2$, eventually approaching levels seen with the intact endothelium before (11, 59-61) (Figure 1).

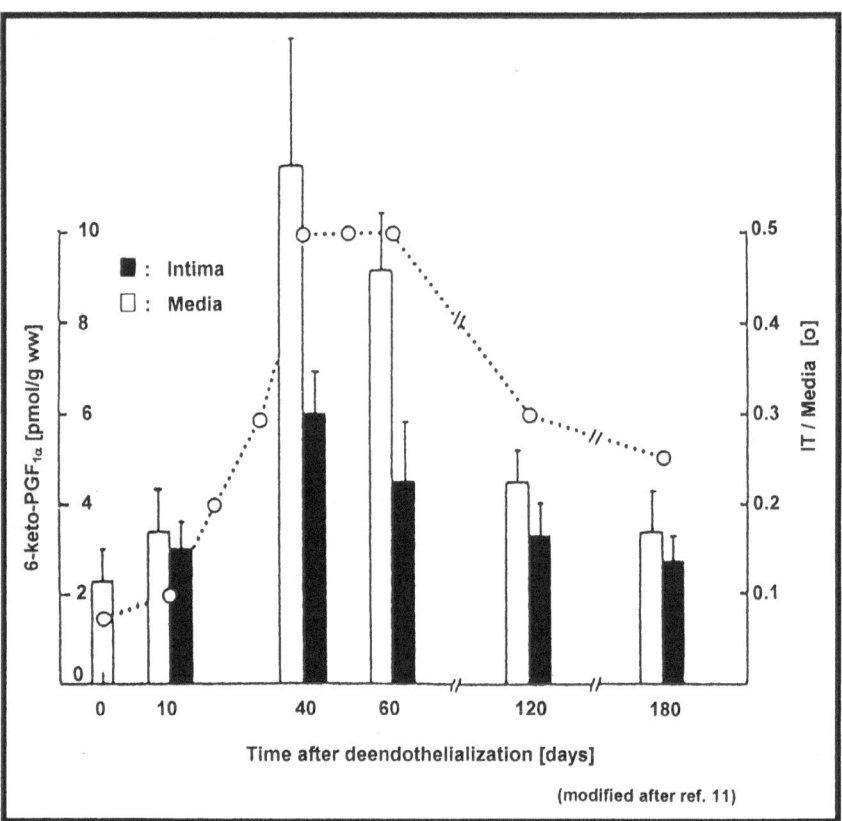

Figure 1. Time-dependent changes in PGI$_2$ production (6-keto-PGF$_{1\alpha}$) by tissue samples from intimal thickening (■) and media (□) of the deendothelialized rabbit aorta. O: indicates the intima/media ratio (modified after ref. [11])

Additional sources of prostaglandin formation by smooth muscle cells

Endothelial injury is associated with the expression of adhesion molecules allowing for close and sustained interactions between the vessel wall and blood constituents such as platelets. This facilitates transcellular metabolism, specifically in the presence of hyperreactive platelets. It is interesting that PGI_2, even at concentrations that completely block all other platelet functions, does not significantly inhibit platelet adhesion to the subendothelium (62). Thus, platelets, adhering to the injured vessel wall within minutes after injury (63-64) and to atherosclerotic plaques (65), might provide precursor endoperoxides for smooth muscle cell PGI_2 production (66-68). The role of platelets and platelet-derived factors in early mitogenesis is uncertain (69). Clinical atherosclerosis is also typically accompanied by enhanced circulating PGI_2 levels which are thought to be due to platelet-dependent precursor transfer (70-71) to vascular cells. The effects of this transcelluar eicosanoid metabolism on neointima formation are unknown.

Signal transduction pathways of prostaglandins with relevance to their antimitogenic action

PGI_2 (IP) receptors

The biological actions of prostaglandins are mediated through various types of transmembrane G-protein-coupled receptors. This also includes vascular smooth muscle cells (72). Recent in situ hybridization studies of IP-receptor mRNA expression have shown that hybridization signals within several organs were confined to the vasculature and here to vascular smooth muscle cells in arteries of various sizes. Interestingly, these studies also showed that the same receptor molecule was expressed in arterial smooth muscle cells and platelet precursor cells, i.e. the megakaryocytes (73). There is, however, little information of how these receptors are regulated. Dorn and colleagues (74) have shown a down-regulation of thromboxane receptors in PDGF-stimulated vascular smooth muscle cells (74). Earlier in vitro studies have shown that the vasodilatory (75) and antiplatelet (76-78) actions of iloprost are subject to down-regulation of prostaglandin receptors taking place at the level of G_s/adenylate cyclase coupling. In previous studies, we have shown that PGI_2 receptors undergo agonist-induced down-regulation in endothelial cells (79). This raises the possibility that PGI_2 receptors on vascular smooth muscle cells behave similarly. This issue appears not to have been addressed systematically for vascular smooth muscle cells.

PGI₂ and PGE₁ receptors in vascular smooth muscle cells might be different

We have used bovine coronary artery smooth muscle cells (80-82) to analyze the effects of vasodilatory prostaglandins on mitogenesis and to detect possible differences between PGI_2 and PGE_1 receptors (80-81). Smooth muscle cells were prepared and cultured as described (82). Smooth muscle cells of passages 4-6 were seeded in 24-well plates (5×10^4 cells/well) and cultivated until subconfluency for 72 h. Serum was removed for the following 24 h in order to stop cell proliferation. The incubation medium was supplemented with 3 μM indomethacin to prevent endogenous prostaglandin biosynthesis. Without indomethacin a 6-7-fold increase in PGI_2-production was found (83). DNA synthesis was assayed by immunofluorescence labelling of 5-bromo-2'-deoxyuridine (BrdU) and [³H]-thymidine incorporation. Addition of PDGF-BB (20 ng/ml) for 24 h resulted in a significant increase in DNA synthesis, BrdU labelling and increase in cell number at 72 h after addition of the growth factor.

Direct stimulation of adenylate cyclase by forskolin completely suppressed PDGF-BB-induced mitogenesis. Similar results were obtained, when forskolin was added at the G_o/G_1 phase or during the S-phase of the cell cycle. This indicated that DNA synthesis in these cells was regulated by cAMP-dependent mechanisms and that stimulation of cAMP formation inhibited smooth muscle cell-proliferation independent of the cell cycle. Antimitogenic effects were also obtained after addition of the prostacyclin mimetic, iloprost, to these cells. Interestingly, an inhibitory action of iloprost was only obtained after addition during the S-phase of the cell cycle. No inhibition was seen, when the compound was already present at the beginning of growth stimulation. However, forskolin was still active in iloprost-desensitized smooth muscle cells, suggesting a mechanism of desensitization upstream of the adenylate cyclase (Figure 2).

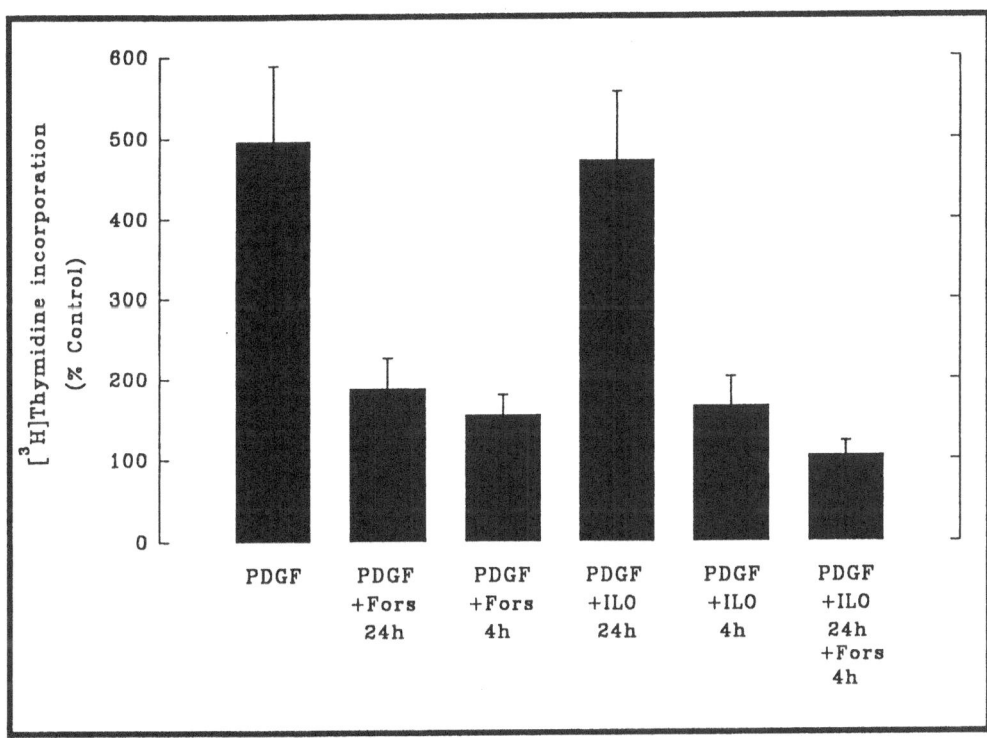

Figure 2. Inhibition of PDGF-induced [³H]-thymidine incorporation by iloprost (100 nM) and forskolin (30 nM). Further explanation see text (adapted from data in ref. [80])

A different picture was seen when PGE_1 was used as putative antimitogen under otherwise identical conditions. At a comparably molar potency (EC_{50}: 3-5 nM), the inhibition of DNA synthesis by PGE_1 was not dependent on the time-point of prostaglandin addition. This suggests that iloprost and PGE_1 act differently on DNA synthesis and that the antimitogenic effect of iloprost but not of PGE_1 was subject of desensitization. To further study this issue, a number of cross-reactivity experiments was conducted. The main finding was that PGE_1 was also active in iloprost desensitized cells as was forskolin. This suggests that different receptor subtypes mediate the antimitogenic action of PGE_1 and PGI_2 and indicates homologous desensitization to PGI_2. It also confirmed that desensitization occured at a level upstream to adenylate cyclase. This finding was different from data in non-vascular cells (NG108-15) where PGE_1 pretreatment produced a 75-80% decrease in IP-receptor number (84).

As discussed before, PGI_2 production in proliferating cells is much higher than in non-proliferating or confluent cells (11, 30, 45). This may explain why other investigators failed to detect inhibition of mitogen-induced growth responses in vascular smooth muscle cells, treated with PGI_2 for 1-4 days (85). In one study with exogenous PGI_2, it was found that only addition of the compound at 10-16 h after growth stimulation of quiescent cells inhibited DNA synthesis (51). This could have been due to the instability of PGI_2 because a stable PGI_2 analogue was more effective in this study. Others demonstrated an inhibition of growth factor-stimulated DNA synthesis in bovine aortic smooth muscle cells by ciprostene. There was a maximum inhibition of DNA synthesis (by 50%) and an increase in cAMP when the compound was added during the first 3 h of growth factor stimulation of previously quiescent cells (23). Alternatively, inhibition of PGI_2 generation, as done in the present experiments by indomethacin, might result not only in the detection of growth inhibitory potential of endogenous PGI_2 (51) but also a hypersensitivity to the agonist. In this context, the low IC_{50}, amounting to ≤ 5 nM, is remarkable: in most of the other studies without inhibition of endogenous prostaglandin formation, the IC_{50} was higher than 1 µM (20, 51).

Intracellular signalling

Receptors for vasodilatory prostaglandins are coupled to different intracellular signalling cascades via different G-proteins. At least three different transduction systems are involved: G_s- or G_i-coupled control of adenylate cyclase activity, G_q-coupled activation of phospholipase C which induces phospholipid breakdown and generates the signal molecules IP_3 and diacylglycerol and results in Ca^{++}-mobilisation (86-87). Opening of ATP-dependent potassium channels appears to cause prostacyclin-induced vasodilatation (88). So far, no clear relationships to mitogenic reactions have been described.

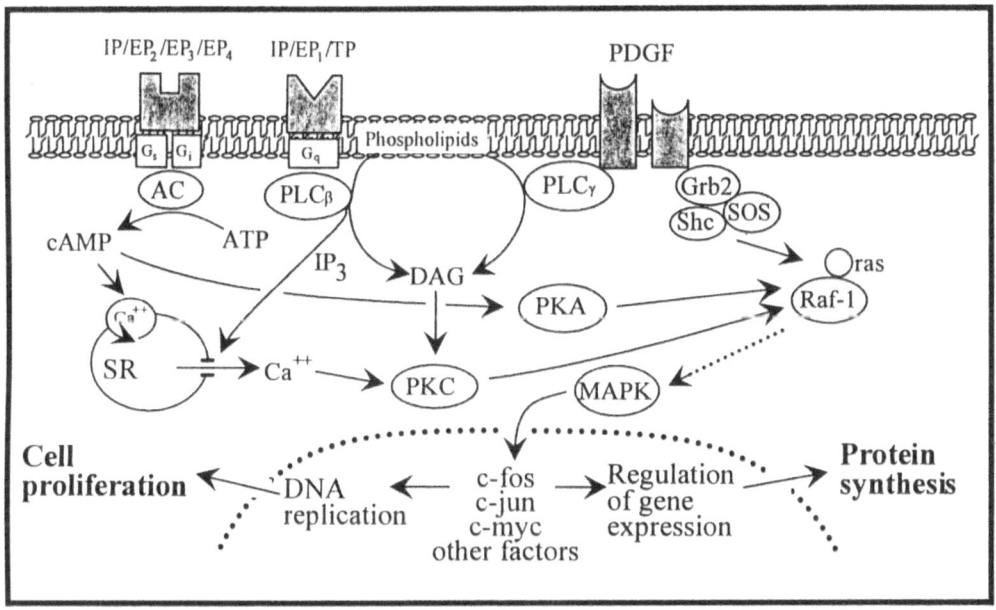

Figure 3. Major signal transduction pathways in vascular smooth muscle cells (SMC). Further explanation see text.

Cyclic AMP

Many of the vascular cell and blood cell actions of PGI_2 are due to G_s-coupled adenylate cyclase stimulation. After the original study by Stout (89), cAMP was considered in numerous follow-up investigations (21, 90-91) to be also the messenger of its growth modulating effects. However, the situation is probably more complex, because intracellular cAMP serves different functions during the cell cycle (92). Franks and colleagues (93) have shown that adenylate cyclase activity and DNA synthesis are stimulated in parallel by growth factors in vascular smooth muscle cells. Cycloheximide blocked DNA synthesis but not the increase in adenylate cyclase activity, suggesting that DNA replication but not adenylate cyclase stimulation required de novo protein synthesis. Different results between quiescent and cycling smooth muscle cells were also obtained by Owen (94). She showed that a (transient) increase in cAMP formation was necessary to initiate the cell cycle while increased cAMP during the S-phase had the opposite effect and inhibited DNA synthesis. However, her studies were performed in fetal cells which might differ from those, prepared from adult animals or humans. Loesberg and colleagues demonstrated that PGE_1 inhibited PDGF-induced DNA synthesis when added during G_1 or at the beginning of S-phase (95). These experiments were conducted in human adult aortic smooth muscle cells. Another study reported an inhibition of growth factor-stimulated DNA synthesis in bovine aortic smooth muscle cells by ciprostene. There was a maximum inhibition of DNA synthesis (by 50%) and an increase in cAMP when the compound was added during the first 3 h of growth factor stimulation (23). In some studies, phosphodiesterase inhibitors did not cause cAMP accumulation or inhibition of DNA synthesis (51), whereas in others they did (14).

This suggests that a rise in intracellular cAMP during G_o or G_1 might be necessary for entering the cell cycle and its progression while elevations of cAMP in late G_1 or early S phase will inhibit proliferation. The first event may be important for wound healing, including stimulation of cell division, the second for the role of prostacyclins as antiproliferative agents in atherosclerotic plaque formation (93-94).

Phospholipase C and protein kinase C (PKC)

Another signalling pathway, related to mitogenic stimuli, is the G-protein-coupled, phospholipase C (PLC)-dependent phospholipid breakdown with generation of the signalling molecules IP_3 and diacycylglycerol and subsequent stimulation of cytosolic Ca^{++} and protein kinase C (PKC) (Figure 2). In several non-vascular cells, IP receptors have found to be coupled to PLC-dependent pathways (86, 96), raising the possibility that these pathways might also be involved in mitogenic signalling in vascular smooth muscle cells. Clearly, EP_1 receptors, stimulated for example by PGE_1, will also cause PLC activation (87). Fukuo and colleagues (97) have recently described that activation of PKC by phorbol ester enhanced prostacyclin (OP-41483)-induced cAMP accumulation in vascular smooth muscle cells and at the same time enhanced its antimitogenic effect. These actions were no longer detectable after desensitization of PKC, suggesting that the enhancement of OP-41483-induced cAMP accumulation by elastase and, eventually, its antimitogenic effect requires activation of PKC. However, PKC(s) have been shown to exert both proliferative and antiproliferative effects on cultured vascular smooth muscle cells (98) and the involvement of Ca^{++}-activated PKC(s) in control of proliferation also depends on the type of mitogen. Thus, any definite conclusions on the possible involvement of PKC in prostaglandin-induced modulation of mitogenesis are difficult.

Distal signalling

Growth factors, such as PDGF, generally act via tyrosine kinase receptors that are coupled to the mitogen-activated protein (MAP) kinase cascade as the major intracellular signal transduction pathway (99). PDGF is also generated by smooth muscle cells themselves after injury (100). Stimulation of MAP kinase activity by growth factors follows activation of the ras/Raf pathway (Figure 3). Several recent studies have shown cAMP-dependent phosphorylation of Raf-1 by protein kinase A and subsequent inhibition of the MAP kinase pathway to the cell nucleus (101-104). This raises the possibility that vasodilatory prostaglandins might inhibit ras by cAMP-dependent phosphorylation of Raf-1. In addition, Raf-1 might be the molecule that connects the signalling cascade of G-protein-coupled receptors, such as the IP- or EP-receptor(s), to MAP kinases, i.e. the signal transduction pathway for growth factors.

More recently, proteins upstream of ras have been detected that stimulate guanine nucleotide exchange at ras (99, 105-107) and might also be subject to PGI_2-induced inhibition of growth response. For example, Grb2 binds to phosphorylated tyrosine kinase receptors via its SH2 domain and recognizes phosphorylated tyrosine residues on Shc proteins. Growth factors stimulate tyrosine phosphorylation at Shc2 (108), connecting growth factors, such as PDGFß-receptors, through recruitment of Grb2 to downstream signalling proteins (109-110). The SH3 domains of Grb2 facilitates stable complex formation with the nucleotide exchange factor SOS. By this mechanism, the Shc/Grb2/SOS complex transduces agonist-induced growth receptor signals to ras and MAP kinase.

Ciprostene treatment elevated cellular cAMP and inhibited Shc/Grb2 complex formation and subsequent MAP kinase activation (111). Preliminary data from our laboratory also had shown that the PGI_2 mimetic iloprost inhibited PDGF-induced MAP kinase activation at concentrations that blocked DNA synthesis (81). Interestingly, the time course of cAMP elevation and inhibition of MAP kinase activation by ciprostene were different in rat aortic smooth muscle cells. Shc/Grb2 complex formation was inhibited with a time course that paralleled MAP kinase inhibition whereas stimulation of cAMP occured earlier, at a time when no inhibition of MAP kinase was seen (111). Thus, there might be additional yet unidentified component(s) upstream to Raf-1 that are involved in prostacyclin-related interruption of MAP kinase signalling (111).

Vasodilatory prostaglandins and generation/action of growth promoting signalling molecules

The central mitogenic factor which appears to be involved in many growth stimulating agonists is PDGF. In contrast to other growth factors PDGF stimulates smooth muscle cells dedifferentiation (112). Vasodilatory prostaglandins have been shown to inhibit PDGF formation (18-19, 85). This led to the conclusion that part of the antimitogenic effect might be due to prevention of PDGF formation.

Another growth factor of considerable interest that might be regulated by prostaglandins is hepatocyte growth factor (HGF). HGF is a mesenchymal-derived pleiotropic factor, possessing mitogenic and morphogenic activities on target cells, suggesting a role in tissue organisation and homeostasis. Vasodilatory prostaglandins, including PGE_1 and the PGI_2 mimetic OP-2507 have been found to stimulate HGF production in human skin fibroblasts and vascular smooth muscle cells by up-regulation of HGF-gene transcription, suggesting that these prostaglandins, generated at a local site of tissue injury, may play a role as inducers of HGF and facilitate tissue regeneration and angiogenesis (113).

Vasodilatory prostaglandins and generation/action of growth inhibitory signalling molecules

Several lines of evidence suggest interactions between the cyclooxygenase and nitric oxide synthase pathways (7). Koide and colleagues (114) have suggested that cyclic AMP-elevating agents stimulate an inducible type of nitric oxide synthase in cultured vascular smooth muscle cells. On the other hand, Salvemini and colleagues (115) have suggested that nitric oxide activates cyclooxygenase enzymes. Since the generation of both mediators is controlled at the transcriptional level by similar factors in vascular smooth muscle cells, i.e. under the influence of cytokines and growth factors (7), it is conceivable to assume that both act synergistically on growth responses. Both cAMP and cGMP not only inhibit contraction but also proliferation of vascular smooth muscle cells (116-117). Consequently, PGI_2 and NO have been found to continuously suppress smooth muscle cell proliferation in arteries in vivo, via cAMP and cGMP, respectively (90). The antimitogenic effect of cAMP or cAMP analogues was considered to be stronger than that of cGMP elevating agents or analogues such as 8-Br-cGMP (21). However, De Meyer and colleagues (118) found an unchanged receptor-independent NO formation and action in terms of vessel relaxation seven days after endothelial injury in the rabbit. At the same time, i.e. after a fully developed neointima was present, the responses to iloprost were enhanced at unchanged reactions to forskolin and isoproterenol, leading to the hypothesis that the number of IP-receptors might have been up-regulated.

Control of mitogenesis of smooth muscle cells by vasodilatory prostaglandins

Stiles and colleagues (119) have originally developed a concept of a two-component model of cell proliferation in fibroblasts. They have classified mitogens into competence factors, e.g. bFGF and PDGF and progression factors, such as somatomedin, both necessary in sequence to initiate cell division. As noted above, the action of vasodilatory prostaglandins is dependent on the cell cycle and might result in an initiation of division at the G_o/G_1 borderline. Mitogenic agonists, such as arginine vasopressin have been shown to inhibit DNA synthesis in vascular smooth muscle by inhibiting progression of G_1 to S phase. This inhibitory action was mediated by prostaglandin synthesis (50). Interestingly, hypoxic human vascular endothelial cells were also found to release prostaglandins as an overall response caused by smooth muscle cells proliferation. These prostaglandins acted synergistically with bFGF by a process that did not require de novo protein synthesis and was sensitive to indomethacin (120). Studies in proliferating vascular smooth muscle cells have also shown that the peak increase in PGI_2 formation initiated by PDGF occured at the G_o/G_1-phase of the cell cycle and might act to inhibit the transition from G_o/G_1 into the synthesis phase of the cell cycle (121).

High endogenous PGI_2 production, for example in acute myocardial ischaemia, is associated with a marked platelet desensitization to iloprost (122). Chronic administration of PGI_2 to patients with advanced atherosclerosis was found to result in a 3-fold decrease of platelet sensitivity against PGI_2 after 3 days of infusion (123). Finally, there was no inhibition of ADP-induced platelet aggregation in patients, treated with intravenous infusions of PGI_2 for 36 h subsequent to PTCA (41).

Conclusions

Vasodilatory prostaglandins have a modulatory effect on proliferation of vascular smooth muscle cells. In general, the effects are antimitogenic in nature and appear to be regulated at the receptor level. In addition, it might be assumed that different sites of action are responsible for antimitogenic effects of prostacyclin as compared to E-type prostaglandins. These differences may involve different receptors at the cell membrane, different signalling pathways, or - hypothetically - direct effects of PGE_1 on the nucleus. The differential succeptibility to tolerance development to PGI_2 as compared to PGE_1 may not only be of relevance for control of smooth muscle cell proliferation by endogenously synthesized prostaglandins but also may have an impact for the clinical use of PGI_2 analogues.

Acknowledgements
The experiments of this study were supported by the Deutsche Forschungsgemeinschaft (SFB 351, D7). The authors are grateful to Erika Lohmann for excellent secretarial assistance.

References

1.   Campbell JH, Campbell GR. Endothelial cell influences on vascular smooth muscle phenotype. Ann Rev Physiol 1986; 48:295-306.

2.   Ross R. The pathogenesis of atherosclerosis - an update. N Engl J Med 1986; 314:488-500.

3.   Bauters C, Meurice T, Hamon M, McFadden E, Lablanche J-M, Bertrand ME. Mechanisms and prevention of restenosis: From experimental models to clinical practice. Cardiovasc Res 1996; 31:835-846.

4.   Campbell JH, Campbell GR. Culture techniques and their applications to studies of vascular smooth muscle. Clin Sci 1993; 85:501-513.

5.   Reines EW, Ross R. Smooth muscle cells and the pathogenesis of the lesions of atherosclerosis. Br Heart J 1993; 69 (suppl):S30-S37.

6.   Venance SL, Bennett BM, Pang SC (1993) Forskolin and isoproterenol effect discrete responses on epidermal growth factor induced DNA synthesis in aortic smooth muscle cells. Can J Physiol Pharmacol 71:800-805.

7.   Wu KK. Inducible cyclooxygenase and nitric oxide synthase. Adv Pharmacol 1995; 33:179-207.

8.   Owens GK. Role of alterations in the differentiated state of vascular smooth muscle cells in atherogenesis. In: Atherosclerosis and Coronary Artery Disease. Fuster V, Ross R, Topol EJ, editors. Philadelphia: Lippincott-Raven, 1996: 401-420.

9.   Schrör K. Prostacyclin (PGI$_2$) and atherosclerosis. In: Endothelium in Clinical Practice: Source and Target of Novel Concepts and Therapies. Rubanyi G, Dzau V, editors. Richmond: Marcel Dekker, 1997: 1-44.

10. Sjolund M, Nilsson J, Palmberg L, Thyberg J. Phenotype modulation of primary cultures of arterial smooth muscle cells. Dual effect of prostaglandin E$_1$. Differentiation 1984; 27:158-162.

11. Larrue J, Daret D, Demond-Henri J, Allières C, Bricaud H. Prostacyclin synthesis by proliferative aortic smooth muscle cells. A kinetic in vivo and in vitro study. Atherosclerosis 1984; 50:63-72.

12.   Pomerantz KB, Hajjar DP. Eicosanoids in regulation of arterial smooth muscle cell phenotype, proliferative capacity, and cholesterol metabolism. Arteriosclerosis 1989; 9:413-429.

13.   Huttner II, Gwebu ET, Panganamala RV, Sharma HM, Geer JC. Fatty acids and their prostaglandin derivatives: inhibitors of proliferation in aortic smooth muscle cells. Science 1977; 197:289-291.

14.   Nilsson J, Olsson AG. Prostaglandin E_1 inhibits DNA synthesis in arterial smooth muscle cells stimulated with platelet-derived growth factor. Atherosclerosis 1984; 53:77-82.

15.   Sinzinger H, Silberbauer K, Winter M, Auerswald W. Is human arterial smooth muscle cell proliferation regulated by prostacyclin? Exp Pathol 1979; 17:354-356.

16.   Pietilä K, Moilanen T, Nikkari T. Prostaglandins enhance the synthesis of glycosaminoglycans and inhibit the growth of rabbit aortic smooth muscle cells in culture. Artery 1980; 7:509-518.

17.   Owen NE. Prostacyclin can inhibit DNA synthesis in vascular smooth muscle cells. In: Prostaglandins, leukotrienes, and lipoxins. Bailey JM, editor. New York: Plenum Press 1985: 193-204.

18.   Sinzinger H, Steurer G, Kaliman J, Ettl K. Interaction between the platelet derived growth factor and prostaglandin I_2 is important for atherosclerosis. Adv Prostaglandins Thrombox Leukotr Res 1987; 17:216-218.

19.   Sinzinger H, Zidek Th, Fitscha P, O'Grady J, Wagner O, Kaliman J. Prostaglandin I_2 reduces activation of human arterial smooth muscle cells in-vivo. Prostaglandins 1987; 33:915-918.

20.   Uehara Y, Ishimitsu T, Kimura K, Ishii M, Ikesa T, Sugimoto T. Regulatory effects of eicosanoids on thymidine uptake by vascular smooth muscle cells of rats. Prostaglandins 1988; 36:847-857.

21.   Southgate K, Newby AC. Serum induced proliferation of rabbit aortic smooth muscle cells from the contractile state is inhibited by 8-Br-cAMP but not 8-Br-cGMP. Atherosclerosis 1990; 82:113-123.

22.   Nakaki T, Ohta M, Kato R. Inhibition by prostacyclin and carbacyclins of endothelin-induced DNA synthesis in cultured vascular smooth muscle cells. Prostaglandins Leukot Essent Fatty Acids 1991; 44:237-239.

23.   Shirotani M, Yui Y, Hattori R, Kawai C. U 61,431F, a stable prostacyclin analogue, inhibits the proliferation of bovine vascular smooth muscle cells with little antiproliferative effect on endothelial cells. Prostaglandins 1991; 41:97-110.

24.   Murase T, Kozawa O, Miwa M, Tokuda H, Kotoyori J, Kondo K, Oiso Y. Regulation of proliferation by vasopressin in aortic smooth muscle cells: function of protein kinase C. J Hypertension 1992; 10:1505-1511.

25.   Koh E, Morimoto S, Jiang B, Inoue T, Nabata T, Kitano S, Yasuda O, Fukuo K, Ogihara T. Effects of beraprost sodium, a stable analogue of prostacyclin, on hyperplasia, hypertrophy and glycosaminoglycan synthesis of rat aortic smooth muscle cells. Artery 1993; 20:242-252.

26.   Asada Y, Kisanuki A, Hatekeyama K, Takahama S, Kurozumi S, Sumiyoshi A. Inhibitory effects of prostacyclin analogue, TFC 132, on aortic neointimal thickening in vivo and smooth muscle cell prolferation in vitro. Prostaglandins Leukotrienes Ess Fatty Acids 1994; 51:245-248.

27.   Pasricha PJ, Hassoun PM, Teufel E, Landman MJ, Fanburg BL. Prostaglandins E_1 and E_2 stimulate the proliferation of pulmonary artery smooth musle cells. Prostaglandins 1992; 43:5-19.

28.   Palmberg L, Lindgren JA, Thyberg J, Claesson HE. On the mechanism of induction of DNA synthesis in cultured arterial smooth muscle cells by leukotrienes. Possible role of prostaglandin endoperoxide synthase products and platelet-derived growth factor. J Cell Sci 1991; 98:141-149.

29.   Hayes LW, Goguen CA, Stevens AL, Magargal SW, Slakey LL. Enzyme activities in endothelial cells and smooth muscle cells from swine aorta. Proc Natl Acad Sci USA 1979; 76:2532-2535.

30.   Hajjar DP, Weksler BB. Modulation of arterial cholesteryl ester metabolism by prostacyclin and prostaglandin E_2. Adv Prostaglandins Thrombox Leukotr Res 1985; 15:249-252.

31.   Braun M, Hohlfeld T, Kienbaum P, Weber A-A, Sarbia M, Schrör K. Antiatherosclerotic effects of oral cicaprost in experimental hypercholesterolemia in rabbits. Atherosclerosis 1993; 103:93-105.

32.   Isogaya M, Yamada N, Koike H, Ueno Y, Kumagai H, Ochi Y, Okazaki S, Nishio S. Inhibition of restenosis by beraprost sodium (a prostaglandin I_2 analogue) in the atherosclerotic rabbit artery after angioplasty. J Cardiovasc Pharmacol 1995; 25:947-952.

33.   Orekhov AN, Tertov VV, Kudryashov AS, Khashimov KA, Smirnov VN. Primary culture of human aortic intima cells as a model for testing antiatherosclerotic drugs. Effects of cyclic AMP, prostaglandins, calcium antagonists, antioxidants and lipid-lowering agents. Atherosclerosis 1986; 60:101-110.

34.   Levitt NA, Dryjski M, Tluczek J, Bjornsson TD. Evaluation of a prostacyclin analog, iloprost, and a thromboxane A_2 receptor antagonist, daltroban, in experimental intimal hyperplasia. Prostaglandins 1991; 41:1-6.

35.   Hehrlein C, Svenson RH, McKusik AP, Chuang CH, Tuntelder JR, Tatsis GP, Littmann L, Thompson M. Thermal laser arterial injury and prostacyclin administration in dogs: thrombotic and hyperplastic consequences. J Cardiovasc Surg 1992; 33: 366-371.

36.   Zoldhelyi P, McNatt J, Xu X-M, Loose-Mitchell D, Meidell RS, Clubb FJ, Buja M, Willerson JT, Wu KK. Prevention of arterial thrombosis by adenovirus-mediated transfer of cyclooxygenase gene. Circulation 1996; 93:10-17.

37.   Löbel P, Schrör K. Stimulation of vascular prostacyclin and inhibition of platelet function by oral defibrotide in cholesterol-fed rabbits. Atherosclerosis 1989; 80: 69-79.

38.   Isaka Y, Handa N, Imaizumi M, Kimura K, Kamada T. Effect of TRK 100, a stable orally active prostacyclin analogue, on platelet function and plaque size in atherothrombotic strokes. Thromb Haemost 1991; 65:344-350.

39.   Darius H, Nixdorff U, Zander J, Rupprecht HJ, Erbel R, Meyer J. Effects of ciprostene on restenosis rate during therapeutic transluminal coronary angioplasty. Agents Actions 1992; 37:305-311.

40.   Knudtson ML, Flintoft VF, Roth DL, Hansen JL, Duff HJ. Effect of short-term prostacyclin administration on restenosis after percutaneous transluminal coronary angioplasty. J Am Coll Cardiol 1990; 15: 691-697.

41.   Gershlick AH, Spriggins D, Davies SW, Syndercombe Court YD, Timmins J, Timmis AD, Rothman MT, Layton C, Balcon R. Failure of epoprostenol (prostacyclin, PGI$_2$) to inhibit platelet aggregation and to prevent restenosis after coronary angioplasty: results of a randomised placebo controlled trial. Br Heart J 1994; 71:7-15.

42.   Coughlin SR, Moskowitz MA, Zetter BR, Antoniades HN, Levine L. Platelet-dependent stimulation of prostacyclin synthesis by platelet-derived growth factor. Nature 1980; 288:600-602.

43.   Siegle I, Nüsing R, Brugger R, Sprenger R, Zecher R, Ullrich V. Characterization of monoclonal antibodies generated against bovine and porcine prostacyclin synthase and quantitation of bovine prostacyclin synthase. FEBS Lett 1994; 347:221-225.

44.   Smith WL. Prostaglandin biosynthesis and its compartmentation in vascular smooth muscle and endothelial cells. Annu Rev Physiol 1986; 48:251-262.

45.   Rimarachin JA, Jacobson JA, Szabo P, Maclouf J, Creminon C, Weksler BB. Regulation of cyclooxygenase-2-expression in aortic smooth muscle cells. Arterioscler Thromb 1994; 14:1021-1031.

46.   Otto JC, Smith WL. Prostaglandin endoperoxide synthases-1 and -2. J Lipid Med 1995; 12:139-156.

47. Nüsing RM, Klein T, Siegle I, Brugger R, Ullrich V. Regulation of prostanoid synthesis in the cardiovascular system. In: Mediators in the Cardiovascular System: Regional Ischemia. Schrör K, Pace-Asciak CR, editors. Basel. Birkhäuser, Agents Actions Suppl 1995; 45:1-10.

48. Uehara Y, Takada S, Hirawa N, Kawabata Y, Nagata T, Numabe A, Hara H, Kudo I, Ikeda T, Inoue K, Sugimoto T, Omata M. De novo synthesis of phospholipase A_2 and prostacyclin production by proliferating rat smooth muscle cells. Prostaglandins 1993; 46:331-346.

49. Habenicht AJR, Salbach P, Goerig M, Zeh W, Janssen-Timmen U, Blattner C, King WC, Glomset JA. The LDL-receptor pathway delivers arachidonic acid for eicosanoid formation in cells, stimulated by platelet-derived growth factor. Nature 1990; 345:634-636.

50. Libby P, Warner SJC, Friedman GB. Interleukin 1: A mitogen for human vascular smooth muscle cells that induces the release of growth inhibitory prostanoids. J Clin Invest 1988; 81:487-498.

51. Morisaki N, Kanzaki T, Motoyama N, Saito Y, Yoshida S. Cell cycle-dependent inhibition of DNA synthesis by prostaglandin I_2 in cultured rabbit aortic smooth muscle cells. Atherosclerosis 1988; 71:165-171.

52. Hara S, Morishita R, Tone Y, Yokoyama C, Inoue H, Kaneda Y, Ogihara T, Tanabe T. Overexpression of prostacyclin synthase inhibits growth of vascular smooth muscle cells. Biochem Biophys Res Comm 1995; 216:862-867.

53. Dembinska-Kiec A, Gryglewska T, Zmuda A, Gryglewski RJ. The generation of prostacyclin by arteries and by the coronary vascular bed is reduced in experimental atherosclerosis in rabbits. Prostaglandins 1977; 14:1025-1034.

54. Zmuda A, Dembinska-Kiec A, Chytkowski A, Gryglewski RJ. Experimental atherosclerosis in rabbits: Platelet aggregation, thromboxane A_2 generation and antiaggregatory potency of prostacyclin. Prostaglandins 1977; 14:1035-1042.

55. Silberbauer K, Sinzinger H, Winter M. Prostacyclin production by vascular smooth muscle cells. Lancet 1978; 1:1356-1357.

56. Sinzinger H, Silberbauer K, Winter M. Effects of experimental atherosclerosis on prostacyclin (PGI_2) generation in arteries of miniature swine. Artery 1979; 5:448-462.

57. Cragg A, Einzig S, Castenada-Zuniga W, Amplatz K, White JG, Rao GHR. Vessel wall arachidonate metabolism after angioplasty: possible mediators of postangioplasty vasospasm. Am J Cardiol 1983; 51:1441-1445.

58. Mattsson E, Brunkwall J, Bergqvist D. Influence of transluminal angioplasty on the prostanoid release from the vessel wall. Eur J Vasc Surg 1990; 4:11-17.

59.   Eldor A, Falcone DJ, Hajjar DP, Minick CR, Weksler BB. Recovery of prostacyclin
       production by the de-endothelialized rabbit aorta. J Clin Invest 1981; 67:735-741.

60.   DeWitt DL, Day J S, Sonnenburg WK, Smith WL. Concentrations of prostaglandin
       endoperoxide synthase and prostaglandin I_2 synthase in the endothelium and smooth
       muscle of bovine aorta. J Clin Invest 1983; 72:1882-1888.

61.   Mattsson E, Brunkwall J, Fält K, Bergqvist D. Vessel repair after balloon angioplasty:
       morphological appearance and prostacyclin synthesising capacity. Eur J Vasc Surg
       1992; 6:585-592.

62.   Weiss HJ, Turitto VT. Prostacyclin (prostaglandin I_2, PGI_2) inhibits platelet adhesion
       and thrombus formation to the subendothelium. Blood 1979; 53:244-250.

63.   Groves HM, Kinlough-Rathbone RL, Richardson M, Moore S, Mustard JF. Platelet
       interaction with damaged rabbit aorta. Lab Invest 1979; 40:194-200.

64.   Wilentz JR, Sanborn TA, Haudenschild C, Valeri CR, Ryan TJ, Faxon DP. Platelet
       accumulation in experimental angioplasty: time course and relation to vascular injury.
       Circulation 1987; 75:636-642.

65.   van Zanten GH, de Graaf S, Slootweg PJ, Heijnen HFG, Connolly TM, de Groot PG,
       Sixma JJ. Increased platelet deposition on atherosclerotic coronary arteries. J Clin
       Invest 1994; 93:615-632.

66.   Tansik RL, Namm DH, White HL. Synthesis of 6-keto-$PGF_{1\alpha}$ by cultured aortic
       smooth muscle cells and stimulation of its formation in a coupled system with platelet
       lysates. Prostaglandins 1978; 15:399-409.

67.   Papp AC, Crowe L, Pettigrew LC, Wu KK. Production of eicosanoids by
       deendothelialized rabbit aorta: interaction between platelets and vascular wall in the
       synthesis of prostacyclin. Thromb Res 1986; 42:549-556.

68.   Hechtman DH, Kroll MH, Gimbrone Jr. MA, Schafer AI. Platelet interaction with
       vascular smooth muscle in synthesis of prostacyclin. Am J Physiol 1991; 260:H1544-
       1551.

69.   Fingerle J, Johnson R, Clowes AW, Majesky MW, Reidy MA. Role of platelets in
       smooth muscle cell proliferation and migration after vascular injury in rat carotid artery.
       Proc Natl Acad Sci USA 1989; 86:8412-8416.

70.   Tremoli E, Socini A, Petroni A, Galli C. Increased platelet aggregability is associated
       with increased prostacyclin production by vessel walls in hypercholesterolemic rabbits.
       Prostaglandins 1982; 24:397-404.

71.   FitzGerald GA, Smith B, Pedersen AK, Brash AR. Increased prostacyclin biosynthesis
       in patients with severe atherosclerosis and platelet activation. N Engl J Med 1984;
       310:1065-1068.

72. Rücker W, Schrör K. Evidence for high-affinity prostacyclin binding sites in vascular tissue: Radioligand studies with a chemically stable analogue. Biochem Pharmacol 1983; 32:2405-2410.

73. Oida H, Namba T, Sugimoto Y, Ushikubi F, Ohishi H, Ichikawa A, Narumiya S. In situ hybridization studies of prostacyclin receptor mRNA expression in various mouse organs. Br J Pharmacol 1995; 116:2828-2837.

74. Dorn GW II, Becker MW. Growth factors downregulate vascular smooth muscle thromboxane receptors independent of cell growth. Am J Physiol Cell Physiol 1992; 262:C927-C933.

75. Vermue NA, Houwertjes MC. Vasodilatation and receptor desensitization in capillaries of the rabbit ear due to prostaglandin E_1, prostacyclin and ZK 36.374 stimulation. In: Prostaglandins and Other Eicosanoids in the Cardiovascular System. Schrör K, editor. Basel: Karger, 1985: 273-278.

76. Alt U, Leigh PJ, Wilkins AJ, Morris PK, MacDermot J. Desensitization of iloprost responsiveness in human platelets follows prolonged exposure to iloprost in vitro. Br J Clin Pharmacol 1986; 22:118-119.

77. Edwards RJ, MacDermot J, Wilkens AJ. Prostacyclin analogues reduce ADP ribosylation of the α-subunit of the regulator G_s protein and diminish adenosine (A_2) responsiveness of platelets. Br J Pharmacol 1987; 90:501-510.

78. Jaschonek K, Faul C, Schmidt H, Renn W. Desensitization of platelets to iloprost. Loss of specific binding sites and heterologous desensitization of adenylate cyclase. Eur J Pharmacol 1988; 147:187-196.

79. Schröder H, Schrör K. Prostacyclin-dependent cyclic AMP formation in endothelial cells. Naunyn-Schmiedeberg's Arch Pharmacol 1993; 347:101-104.

80. Grosser T, Bönisch D, Zucker T-P, Schrör K. The inhibition of growth factor-stimulated mitogenesis of coronary artery smooth muscle cells by prostacyclin is attenuated by homologous receptor desensitization. Circulation 1994; 90:I-636.

81. Grosser T, Bönisch D, Zucker T-Ph, Schrör K. Iloprost-induced inhibition of proliferation of coronary artery smooth muscle cells is abolished by homologous desensitization. Agents Actions (suppl) 1995; 45:85-91.

82. Grosser T, Zucker T-P, Weber A-A, Schulte K, Sachinidis A, Vetter H, Schrör K. Thromboxane A_2 induces cell signalling but requires platelet-derived growth factor to act as a mitogen. Eur J Pharmacol 1997; 319:327-332.

83. Weber A-A, Zucker T-P, Hasse A, Bönisch D, Wittpoth M, Schrör K. Antimitogenic effects of vasodilatory prostaglandins in coronary artery smooth muscle cells. Basic Res Cardiol 1997 (in press)

84.   Krane A, MacDermot J, Keen M. Desensitization of adenylate cyclase responses following exposure to IP prostanoid receptor agonists. Homologous and heterologous desensitization exhibit the same time course. Biochem Pharmacol 1994; 47:953-959.

85.   Willis AL, Smith DL, Vigo C. Suppression of principal atherosclerotic mechanisms by prostacyclins and other eicosanoids. Prog Lipid Res 1986; 25:645-666.

86.   Oka M, Negishi M, Nishigaki N, Ichikawa A. Two types of prostacyclin receptor coupling to phosphatidylinositol hydrolysis in a cultured mast cell line, Bnu-2c13 cells. Cell Signalling 1993; 5:643-650.

87.   Coleman RA, Smith WL, Narumiya S. VIII. International Union of Pharmacology. Classification of prostanoid receptors: Properties, distribution, and structure of the receptors and their subtypes. Pharmacol Rev 1994; 46:205-229.

88.   Siegel G, Carl A, Adler A, Stock G. Effect of the prostacyclin analogue iloprost on K^+-permeability in the smooth muscle cells of the canine carotid artery. Eicosanoids 1988; 2:213-222.

89.   Stout RW. Cyclic AMP: A potent inhibitor of DNA synthesis in cultured arterial endothelial and smooth muscle cells. Diabetologia 1982; 22:51-55.

90.   Assender JW, Southgate KM, Hallett MB, Newby AC. Inhibition of proliferation, but not of Ca^{2+} mobilization, by cyclic AMP and GMP in rabbit aortic smooth muscle cells. Biochem J 1992; 288:527-532.

91.   Wiley MH, Feingold KR, Grunfeld C, Quesney-Huneuus V, Wu JM. Evidence for cAMP-independent inhibition of S-phase DNA synthesis by prostaglandins. J Biol Chem 1983; 258:491-496.

92.   Boynton AL, Whitfield JF. The role of cyclic AMP in cell proliferation: a critical assessment of the evidence. Adv Cyclic Nucleotide Res 1983; 15:193-294.

93.   Franks DJ, Plamondon J, Hamet P. An increase in adenylate cyclase activity precedes DNA synthesis in cultured vascular smooth muscle cells. J Cell Physiol 1984; 119:41-45.

94.   Owen NE. Effect of prostaglandin E_1 on DNA synthesis in vascular smooth muscle cells. Am J Physiol Cell Physiol 1986; 250:C584-588.

95.   Loesberg C, van-Wijk R, Zandbergen J, van-Aken WG, van-Mourik JA, De Groot PG. Cell cycle-dependent inhibition of human vascular smooth muscle cell proliferation by prostaglandin E_1. Exp Cell Res 1985; 160:117-125.

96.   Schwaner I, Seifert R, Schultz G. The prostacyclin analogues, cicaprost and iloprost, increase cytosolic Ca^{++}-concentration in the human erythroleukemia cell line, HEL, via pertussis toxin-insensitive G-proteins. Eicosanoids 1992; 5:S10-S12.

97.   Fukuo K, Morimoto S, Jiang B, Inoue T, Nabata T, Ogihara T. Elastase enhances cAMP accumulation and the inhibition of DNA synthesis induced by OP-41483, a stable prostacyclin analogue, in vascular smooth muscle cells. Atherosclerosis 1994; 110:111-117.

98.   Lee MW, Severson DL. Signal transduction in vascular smooth muscle: diacyclglycerol second messengers and PKC action. Am J Physiol 1994; 267:C659-C678.

99.   Davis RJ. The mitogen-activated protein kinase signal transduction pathway. J Biol Chem 1993; 268:14553-14556.

100.   Walker LN, Bowen-Pope DF, Ross R, Reidy MA. Production of platelet-derived growth factor-like molecules by cultured arterial smooth muscle cells accompanies proliferation after arterial injury. Proc Natl Acad Sci USA 1986; 83:7311-7315.

101.   Cook SJ, McCormick FM. Inhibition by cAMP of Ras-dependent activation of Raf. Science 1993; 262:1069-1072.

102.   Wu J, Dent P, Jelinek T, Wolfman A, Weber MJ, Sturgill TW. Inhibition of the EGF-activated MAP kinase signaling pathway by adenosine 3',5'-monophosphate. Science 1993; 262:1065-1069.

103.   Sevetson BR, Kong X, Lawrence JC. Increasing cAMP attenuates activation of mitogen-activated protein kinase. Proc Natl Acad Sci USA 1993; 90:10305-10309.

104.   Graves LM, Bornfeld KE, Raines EW, Potts BC, MacDonals SG, Ross R, Krebs EG. Protein kinase A antagonizes platelet-derived growth factor induced signaling by mitogen-activated protein kinase in human arterial smooth muscle cells. Proc Natl Acad Sci USA 1993; 90:10300-10304.

105.   Lowenstein EJ, Daly RJ, Batzer AG, Li W, Margolis B, Lammers R, Ullrich A, Skolnik EY, Bar-Sagi D, Schlessinger J. The SH2 and SH3 domains of Grb2 link receptor tyrosine kinases to ras signaling. Cell 1992; 70:431-442.

106.   Egan SE, Giddings BW, Brooks MW, Buday L, Sizeland AM, Weinberg RA. Association of Sos Ras exchange protein with Grb2 is implicated in tyrosine kinase signal transduction and transformation. Nature 1993; 363:45-51.

107.   Rozakis-Adcock M, Fernley R, Wade J, Pawson T, Bowtell D. The SH2 and SDH3 domains of mammalian Grb2 couple the EGF receptor to the ras activator mSos1. Nature 1993; 363:83-85.

108.   Pelicci G, Lanfrancone L, Grignani F, McGlade J, Cavallo F, Forni G, Nicoletti I, Pawson T, Pelici PG. A novel transforming protein (SHC) with an SH2 domain is implicated in mitogenic signal transduction. Cell 1992; 70:93-104.

109. Yokote K, Mori S, Hansen K, McGlade J, Pawson T, Heldin C-H, Calsesson-Welsh, L. Direct interaction between Shc and platelet-derived growth factor beta-receptor. J Biol Chem 1994; 269:15337-15343.

110. Benjamin CW, Jones DA. Platelet-derived growth factor stimulates growth factor receptor-binding protein-2 association with Shc in vascular smooth muscle cells. J Biol Chem 1994; 269:30911-30916.

111. Jones DA, Benjamin CW, Linseman DA. Activation of thromboxane and prostacyclin receptors elicits opposing effects on vascular smooth muscle cell growth and mitogen-activated protein kinase signaling cascades. Mol Pharmacol 1995; 48:890-896.

112. Holycross BJ, Blank RS, Thompson MM, Peach MJ, Owens GK. Platelet-derived growth factor-BB-induced suppression of smooth muscle cell differentiation. Circ Res 1992; 71:1525-1532.

113. Matsumoto K, Okazaki H, Nakamura T. Novel function of prostaglandins as inducers of gene expression of HGF and putative mediators of tissue regeneration. J Biochem 1995; 177:458-464.

114. Koide M, Kawahara Y, Nakayama I, Tsuda T, Yokoyama M. Cyclic AMP-elevating agents induce an inducible type of nitric oxide synthase in cultured vascular smooth muscle cells. J Biol Chem 1993; 268:24959-24966.

115. Salvemini D, Misko TP, Masferrer JL, Seibert K, Currie MG, Needleman P. Nitric oxide activates cycooxygenase enzymes. Proc Natl Acad Sci USA 1993; 90:7240-7244.

116. Fukumoto Y, Kawahara Y, Kariya K, Araki S, Fukuzaki H, Takai Y. Independent inhibition of DNA synthesis by protein kinase C, cyclic AMP and interferon α/ß in rabbit aortic smooth muscle cells. Biochem Biophys Res Comm 1988; 157:337-345.

117. Garg UC, Hassid A. Nitric oxide generating vasodilators and 8-bromo-cyclic guanosine monophosphate inhibit mitogenesis and proliferation of cultured rat vascular smooth muscle cells. J Clin Invest 1989; 83:1774-1777.

118. De Meyer GRY, Bult H, Van Hoydonck A-E, Jordaens FH, Buyssens N, Herman AG. Neointima formation impairs endothelial muscarinic receptors while enhancing prostacyclin-mediated responses in the rabbit carotid artery. Circ Res 1991; 68:1669-1680.

119. Stiles CD, Capone GT, Scher CD, Antoniades HN, van Wyk JJ, Pledger WJ. Dual control of cell growth by somatomedin and platelet-derived growth factor. Proc Natl Acad Sci USA 1979; 76:1279-1283.

120. Michiels C, De Leener F, Arnould T, Dieu M, Remacle J. Hypoxia stimulates human endothelial cells to release smooth muscle cell mitogens; role of prostaglandins and bFGF. Exp Cell Res 1994; 213:43-54.

121. Uehara Y, Numabe A, Kawabata Y, Nagata T, Hirawa N, Ishimitsu T, Matsuoka H, Ikeda T, Sugimoto T. Rapid smooth muscle cell growth and endogenous prostaglandin system in spontaneously hypertensive rats. Am J Hypertension 1991; 4:806-814.

122. Jaschonek K, Karsch KR, Weisenberger H, Tidow S, Faul C, Renn W. Platelet prostacyclin binding in coronary artery disease. J Am Coll Cardiol 1986; 8:259-266.

123. Sinzinger H, Silberbauer K, Horsch AK, Gall A. Decreased sensitivity of human platelets to PGI_2 during long-term intraarterial prostacyclin infusion in patients with peripheral vascular disease - a rebound phenomenon? Prostaglandins 1981; 21:49-51.

AAS 48
Prostaglandins and Control of Vascular
Smooth Muscle Cell Proliferation
© 1997 Birkhäuser Verlag Basel

Antimitotic actions of vasodilatory prostaglandins - clinical aspects

H. Sinzinger, P. Fitscha and H. Kritz

Wilhelm Auerswald Atherosclerosis Research Group (ASF), Nadlergasse 1, A-1090 Vienna and Department of Nuclear Medicine, University of Vienna, Währinger Gürtel 18 - 20, A-1090 Vienna, Austria

Summary. A variety of in-vitro antiatherosclerotic actions, among them those on vascular smooth muscle cells (mitotic activity, proliferation, extracellular matrix production), have been identified especially for PGE_1 and PGI_2, and proven in experimental animals. Ex-vivo data in humans are not yet available. We examined the effect of PGE_1-, PGI_2- and iloprost therapy of various duration (1 - 4 weeks) on smooth muscle cells (mitosis, proliferation, prostaglandin formation from exogenous and endogenous substrate) derived from vascular surgery samples. In-vivo PG-therapy decreases $[^3H]$-thymidine incorporation as well as $[^{35}]S$- and $[^{14}C]$-proline uptake. These effects are dependent on the duration of treatment, PGE_1 being trendwise more effective. Arachidonic acid conversion to PGI_2 is significantly enhanced in activated smooth muscle cells of the plaque, both in the intima as well as in the media. Due to the activation of the gene for COX-2, the actual synthesis of PGI_2 as well as the conversion rate to 6-oxo-$PGF_{1\alpha}$ are increased in activated smooth muscle cells, an effect being abolished by the PG`s administered. It can thus be concluded that PG-therapy for advanced atherosclerosis seems to affect vascular smooth muscle cells beneficially, decreasing mitotic and proliferative activity as well as collagen and glycosaminoglycan synthesis. The somewhat less pronounced effect for PGI_2 and iloprost could be explained by desensitization at the receptor level as preliminary findings suggest. This could become even more relevant if a long-term administrable stable (oral) analogue becomes available for routine therapy.

Introduction

Besides a great variety of antiatherosclerotic actions on various systems, the antiaggregatory prostaglandins (PG) have been shown to exert a key regulatory role on vascular smooth muscle cell (SMC) activity and function. The antiaggregatory PGE_1, PGI_2 and their respective analogues have been shown to exert antimitotic and antiproliferative actions in-vitro (1, 2, 3, 4) and in various experimental models (5).

PGE_1 has the ability to inhibit PDGF-induced DNA-synthesis in rat arterial SMCs (6). In parallel PGE_1 and PGI_2 are inhibiting the release of platelet α-granule products including PDGF (7), thus indirectly exerting an antiproliferative action (8, 9). Loesberg (10) et al. reported for the first time that the effect of PGE_1 on SMC-proliferation is cell-cycle dependent. In their experiments, the inhibitory action of PGE_1 on human SMC-proliferation occurred primarily in G1 phase with no effect in S phase. In contrast, PGE_1 increased SMCs' DNA-synthesis (G0/G1 phase).

PGI_2 is the major cyclooxygenase (COX) product synthetized by arterial SMC. After the finding that SMC generate relevant amounts of PGI_2 (11, 12), we described for the first time that the PGI_2-synthesis by activated (A) SMC is significantly enhanced (13). The effects of PGE_1 and PGI_2 on SMC are due to an elevation of intracellular cAMP (14). Interestingly, this has been shown in the intima and in the media as well. This enhanced PGI_2-synthesis could result in a stimulation of intracellular cAMP and, thereby, an inhibition of further proliferation and interruption of a vitious cycle (15). Prolonged activation of SMC and increased local PGI_2-availability, however, could cause a down-regulation of the respective PG-receptors. In addition to the normal baseline PG-production via the constitutive housekeeping enzyme COX-1, various atherogenic stimuli like thrombin, cytokines, growth factors and others (9) are inducing a vastly enhanced PG-formation via COX-2 under cell culture conditions (16). It has been shown, that PGE_1 down-regulates SMC DNA-synthesis in a concentration-dependent manner. This action is influenced by cell-cycle kinetics. PGE_1 exerted an inhibitory action on SMC DNA-synthesis cultured in low serum relative to cycling SMCs maintained in a higher serum concentration.

Many years ago, we started to elaborate the effect of PG-therapy on SMC-proliferation in-vivo by examining morphologically and biochemically vascular tissue samples from patients having been treated before (15, 17, 18). It was the aim of this paper to summarize our findings we obtained over a period of more than 10 years examining mitotic activity, SMC-proliferation, extracellular matrix- and PG-synthesis after PG-therapy (PGE_1, PGI_2, iloprost as compared to untreated controls).

Material and Methods

Surgical material derived from human femoral and popliteal arteries was obtained from patients having been infused before with

a) PGE_1 (Prostavasin®, Schwarz Pharma, Monheim, Germany) at a dose of 5 ng/kg/min for 6 hours a day during 1 to 4 weeks;

b) PGI_2 (Epoprostenol, Flolan®, Welcome Research Labs, Beckenham, Kent/UK) 5 ng/kg/min for 6 hours for 1 to 4 weeks;

c)   Iloprost (Schering, Berlin, Germany) 1 ng/kg/min for 6 hours daily for 1 to 4 weeks.

For morphometric examination tissue samples were fixed immediately in phosphate buffered (pH 7.4) glutaraldehyde, semithin sections were stained using PAS-toluidine blue. The number of ASMC in the intima as well as in the media was quantified and expressed as percentage of total number of SMCs as described originally by us (19). Interobserver variability of morphometric analysis amounted to $4.1 \pm 1.7\%$, intraobserver variation to $2.8 \pm 0.6\%$.

Native samples were used for radioactive precursor incorporation studies. Tissue samples in the absence of growth factors (serum) were incubated in the presence of 5 mg/ml acetylsalicylic acid in order to block PG-synthesis and to avoid artificial in-vitro influences via eventual receptor regulation. $[^3H]$-Thymidine (10 µCi; spec. activity 2 µCi/mmol, Amersham, UK) incorporation as well as $[^{35}S]$- and $[^{14}C]$-proline uptake were quantified after incubation and subsequent washing by autoradiography as a measure for mitotic activity, glycosaminoglycan- and collagen-synthesis. Autoradiography of prestained (haematoxylin-eosin) sections was done using a Kodak NTB 2 emulsion and an exposure time between 2 and 3 weeks. Interobserver variability was $2.1 \pm 0.4\%$, intraobserver variability $1.5 \pm 0.2\%$.

PGI_2-synthesis was determined using the aggregation inhibitory action of PGI_2 on platelets as described originally by Moncada (20). After incubating the tissue samples derived from microdissection for 3 minutes at 37°C in buffer (pH 7.4), the inhibitory capacity of the incubation fluid on platelet aggregation in a Born-type (600 µl platelet rich plasma samples, constant stirring) aggregometer was quantified using a synthetic standard (Cayman Chemicals, Ann Arbor, Mi, USA). Values are given in pg PGI_2/mg tissue wet weight/min. Intraassay variability was $3.9 \pm 1.4\%$, interassay variability $7.2 \pm 2.5\%$. [^{14}C]-Arachidonic acid (AA) conversion rate (in %) was determined by means of radiothinlayer chromatography in comparison to radiolabeled standards (Amersham, Buckinghamshire, UK).

Statistical analysis

The values are given as mean $\pm$ SD; calculation for significance was done using Student's t-test. A p-value of < 0.01 was considered as significant.

Results

In human atherosclerotic lesions activation of SMC is an early event. They are characterized by their typical phenotype and can easily be identified using PAS-toluidine blue staining by their chromophilia.The percentage of ASMC in lesions of younger people is higher as compared to that derived from older ones (Table 1).

Table 1. Distribution of activated smooth muscle cells (ASMC) in human arteries

Area	41-50a	n	51-60a	n	61-70a	n	Therapy
Intima	29 ± 9		27 ± 7		19 ± 4		---
Media	9 ± 4	3	7 ± 3	9	4 ± 3	8	---
Intima	$17 \pm 6^*$		$14 \pm 7^*$		$10 \pm 3^*$		PGI_2
Media	$5 \pm 2^*$	4	4 ± 2	7	4 ± 2	6	PGI_2
Intima	$15 \pm 7^*$		$16 \pm 6^*$		$11 \pm 3^*$		PGE_1
Media	$4 \pm 2^*$	4	5 ± 2	7	4 ± 2	7	PGE_1
Intima	18 ± 5		17 ± 4		12 ± 4		Iloprost
Media	5 ± 2	5	6 ± 2	4	5 ± 2	4	Iloprost

$x \pm SD$; % of total SMC; $^* p < 0.01$; a = age in years

Furthermore, ASMC are more densely packed in the plaque intima and much less in the adjacent media. PG-therapy with either of the compounds administered results in a decrease of ASMC being by far more pronounced in the intima. The effect of therapy clearly is dependent on the duration of therapy (Table 2) reaching significance, however, as early as after only 1 week.

Table 2. Percentage of activated smooth muscle cells (ASMC) in the plaque - Influence of PG-therapy

						weeks				
PG	n	Co	n	1	n	2	n	3	n	4
PGE_1				$18 \pm 3^*$		$13 \pm 3^*$		$11 \pm 3^*$		$10 \pm 3^*$
			6		4		4		3	
PGI_2		29 ± 4		$18 \pm 3^*$		$15 \pm 4^*$		$14 \pm 5^*$		$15 \pm 4^*$
	15		5		6		3		5	
Iloprost				$17 \pm 4^*$		$15 \pm 4^*$		$15 \pm 3^*$		$14 \pm 2^*$
			4		4		4		5	

age group 51 - 60 years
$x \pm SD$; values in %; $^* p < 0.01$

Interestingly, after 2 weeks of treatment, in the PGE$_1$-treated group the ASMC-content is further decreasing, while in the patients treated with PGI$_2$ or its stable analogue iloprost, their number is plateauing. The mitotic activity as determined by [^{3}H]-thymidine incorporation (Table 3) exhibits a decrease, dependent on the duration of PG-application. Due to the higher variations, only after 4 weeks and only in the PGE$_1$-group, however, the values reached the level of significance.

Table 3. Effect of PG-treatment on [^{3}H]-thymidine incorporation in-vitro

| PG | Co | weeks | | | |
		1	2	3	4
PGE$_1$		0.42 ± 0.10	0.40 ± 0.05	0.36 ± 0.05	$0.28 \pm 0.06*$
n		8	6	4	8
PGI$_2$	0.48 ± 0.12	0.40 ± 0.09	0.41 ± 0.08	0.32 ± 0.06	0.33 ± 0.07
n	15	4	5	6	6
iloprost		0.41 ± 0.07	0.38 ± 0.09	0.37 ± 0.11	0.34 ± 0.10
n		6	5	8	7

[^{3}H]-thymidine uptake in %; x $\pm$ SD; * $p < 0.01$

In parallel, the precursor incorporation is decreased, indicating a lower extracellular matrix production induced by PG-therapy (Table 4,5). [^{14}C]-Proline incorporated in-vitro as an indicator of collagen synthesis decreases, again PGE$_1$ having a more pronounced effect (Table 4). The effect is clearly time-dependent, reaching the level of significance after 2 weeks of therapy. The incorporation of [^{35}S] as an indicator of glycosaminoglycan synthesis (Table 5) shows an almost identical behaviour, the improvement becoming significant, however, about 1 week later.

Table 4. Decreased $[^{14}C]$-proline incorporation in-vitro after PG-therapy

PG	Co	1	2	3	4
			weeks		
PGE_1		397 ± 21	$351 \pm 34*$	$328 \pm 40*$	$302 \pm 39*$
n		8	6	6	8
PGI_2	436 ± 43	395 ± 36	$368 \pm 41*$	$360 \pm 39*$	$345 \pm 31*$
n	15	4	6	7	6
iloprost		391 ± 42	$372 \pm 37*$	$366 \pm 41*$	$350 \pm 29*$
n		6	5	8	7

$x \pm SD$; values in cpm/mg protein; $* p < 0.01$

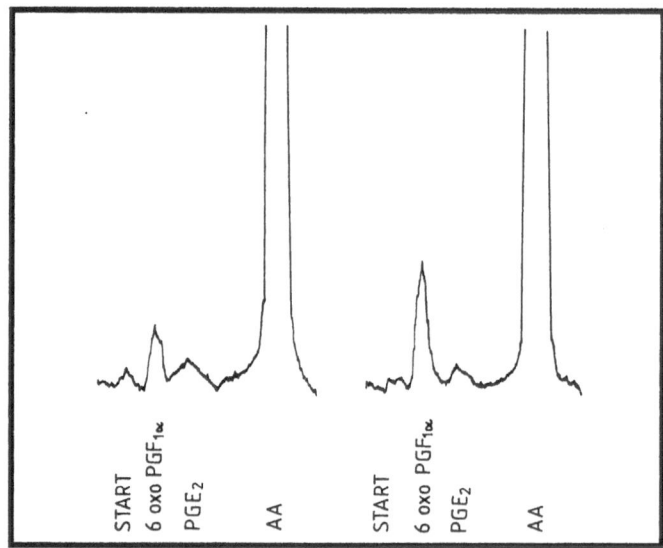

Figure 1. Influence of PG-therapy on conversion of $[^{14}C]$-AA by radiochromatography. The left radiochromatogram shows a peak for 6-oxo-$PGF_{1\alpha}$ (1.07%) and PGE_2 (0.26%) only (sample derived from a 55-years aged male non-smoker; popliteal artery, PGE_1-treated for 3 weeks). In the right radiothinlayer chromatography conversion to 6-oxo-$PGF_{1\alpha}$ (1.53%) and PGE_2 (0.29%) is visible. Popliteal artery derived from a 58-years aged non-smoking male not having been treated with PG's. Note the apparent difference in the 6-oxo-$PGF_{1\alpha}$ peak.

Table 5. [^{35}S]-Incorporation in-vitro - Benefit induced by PG-treatment

PG	Co	1	2	3	4
			weeks		
PGE$_1$		486 ± 38	449 ± 36	418 ± 29*	397 ± 28*
n		8	6	6	8
PGI$_2$	524 ± 51	481 ± 42	439 ± 41	447 ± 41*	451 ± 39
n	15	4	6	7	6
iloprost		485 ± 33	466 ± 34	459 ± 38	444 ± 29*
n		6	5	8	7

x ± SD; values in cpm/mg protein; * p < 0.01

Conversion of exogenous [^{14}C]-arachidonic acid (AA) to 6-oxo-PGF$_{1\alpha}$ is significantly higher in ASMC as compared to contractile SMC (CSMC). Besides 6-oxo-PGF$_{1\alpha}$ a notable peak can only be detected for PGE$_2$ (Figure 1), showing a non-significant decline in samples derived from PG-treated as compared to non-treated patients. No difference between ASMC in the media and the intima of lesioned vessels can be detected. PG-therapy results in a non-significant lowering of conversion rate by ASMC in all the 3 treatment groups, while the metabolic profile of CSMC both in intima and media as well seems to be completely unchanged. The data derived for femoral (Table 6) and popliteal (Table 7) arteries are almost identical. This behaviour of exogenous substrate is reflected by PGI$_2$-synthesis (endogenous substrate) determined by bioassay, too.

Table 6. [^{14}C]-AA-conversion of human femoral artery tissue to 6-oxo-PGF$_{1\alpha}$ after PG-therapy

	ASMC	CSMC	n
CO	1.45 ± 0.27*	0.79 ± 0.21	10
PGE$_1$	1.13 ± 0.22	0.75 ± 0.26	7
PGI$_2$	1.14 ± 0.24	0.73 ± 0.22	4
iloprost	1.11 ± 0.20	0.72 ± 0.19	8

x ± SD; conversion in %; * p < 0.01 (vs. CSMC)

Table 7. Effect of PG-therapy on $[^{14}C]$-AA conversion of human popliteal arterial tissue to 6-oxo-$PGF_{1\alpha}$

	ASMC	CSMC	n
CO	1.48 ± 0.33*	0.76 ± 0.22	12
PGE_1	1.11 ± 0.24	0.72 ± 0.18	8
PGI_2	1.13 ± 0.26	0.70 ± 0.14	6
iloprost	1.09 ± 0.30	0.74 ± 0.19	7

x ± SD; duration of treatment 4 weeks; conversion in %; * p < 0.01

The PGI_2-production of ASMC after PG-therapy tends to be diminished, while that of CSMC is not affected (Tables 8, 9). In a small number of samples examined so far, we saw that this effect becomes visible after one week of therapy only. Longer therapy for up to 4 weeks did not induce a further alteration of the mean values. More material has to be examined to assess these kinetics more in detail.

Table 8. Role of PG-therapy on PGI_2-production of femoral artery SMC

	ASMC	CSMC	n
CO	16.41 ± 3.28*	7.71 ± 2.32	10
PGE_1	10.94 ± 2.91	7.49 ± 2.41	7
PGI_2	11.22 ± 3.16	7.54 ± 1.98	4
iloprost	11.35 ± 3.34	7.55 ± 2.08	8

x ± SD; PGI_2 in pg/mg/min; * p < 0.01 (vs. CSMC)

Table 9. Popliteal artery SMC - PGI_2-production - Influence of PG-treatment

	ASMC	CSMC	n
CO	15.77 ± 3.91*	7..83 ± 2.44	12
PGE_1	11.10 ± 3.21	7.39 ± 2.29	8
PGI_2	10.97 ± 3.56	7.44 ± 2.08	6
iloprost	11.21 ± 2.84	7.48 ± 1.92	7

x ± SD; PGI_2 in pg/mg/min; * p < 0.01 (vs. CSMC)

Conversion of $[^{14}C]$-AA as well as actual PGI_2-synthesis tended to decrease with age in the vascular tissue samples. This was seen in CSMC and ASMC as well.

Discussion

SMC-proliferation is one the central events during human atherogenesis (21). Initially, SMC are changing their phenotype from the contractile to the metabolically active stage being paralleled by a decrease in myofilaments and micropinocytotic vesicles and an increase in chromophilia among others. The rapid de-novo synthesis of COX (22, 23), i.e. the activation of the gene for the inducible cyclooxygenase (COX-2) is an early response to injury in vascular SMC (16). The antiproliferative and antimitogenic effects of antiaggregatory PG's are mediated by the activation of adenylyl cyclase and the subsequent production of cAMP, the latter acting via cAMP-dependent protein kinases. While TXA_2 exerts a direct proliferative action, this is shared also by leukotrienes (B_4, C_4, D_4). LTC_4 and LTD_4. These leukotrienes, however, may also act via a stimulation of PGI_2-synthesis in a biphasic way as inhibitors of cellular proliferation. It is certainly of relevance in this context that the precursor fatty acids are modulating the profile of PG's finally formed. In addition, they (linoleic acid, linolenic acid and in particular arachidonic acid) may directly exert an inhibitory action via tyrosine kinase activity. The activation of SMC with the associated change in the phenotype has been described in-vitro and in-vivo as well. These cells accumulate in a characteristic manner at lesion sites (19). This activation includes the expression of various genes, the release of growth-factors, in particular the platelet-derived growth factor (PDGF), enzymes, cytokines, prostaglandins and finally the production of extracellular matrix components. Stimulatory (24, 25) and inhibitory actions (18, 26, 27) of PGE_1 have been reported. As PGE_1 is substantially metabolized during the first lung passage, it is of particular relevance that 13,14-dihydro-PGE_1 (PGE_0), the biologically active in-vivo metabolite of PGE_1, decreases mitotic activity and SMC-proliferation to an almost comparable extent as compared to the parent compoud PGE_1 (28). These earlier experiments performed in rabbits also revealed that administration of the respective prostaglandin before induction of proliferation was more effective as compared to a treatment thereafter (29). Combined pre- and post-treatment did result in a trendwise further improvement, but not in a significant further benefit. $[^3H]$-Thymidine incorporation in-vitro was decreasing dependent on the time of PG-administration reaching with a PGE_1-therapy after four weeks the level of significance. Furthermore, it has been shown, that the other antiatherogenic actions of PGE_1, such as antiplatelet and hypolipidemic, are shared by PGE_0 to an almost comparable extent. Cell culture studies comparing PGE_1 and iloprost (30) revealed a desensitization for iloprost, but none for PGE_1. In comparison, the effect of PGI_2 and the stable analogue iloprost in our experiments was somewhat less pronounced, however, not resulting in a significant difference. The percentage of ASMC was significantly diminished after only one week of PG-therapy, the further decline for PGI_2 and iloprost was comparable, while the one for PGE_1 was even more pronounced, in particular after 3 and 4 weeks of treatment. We also showed, that PGI_2 suppresses mitotic activity (31) and activation of SMC (15, 17). Inhibitory actions on SMC replication have also been reported for PGA_1, PGB_1, PGD_2, PGE_2 and PGJ_2. Earlier studies revealed that PGE_1 and PGI_2 might be of additive benefit inhibiting vascular SMC proliferation via different mechanisms (32). Present data showing a difference between PGE_1 on the one hand and PGI_2 as well as iloprost on the other hand indicate, that after PGE_1-therapy; there is much less, if any, desensitization. [^{14}C]-proline incorporation in-vitro indicated a significant diminution, starting already after the second week with a further decline thereafter; again the effect being highest after PGE_1-therapy. A quite similar finding can be obtained, monitoring the [^{35}S]-incorporation as a precursor for glycosaminoglycan-synthesis, as compared to the prevalues. For these particular experiments it is of pivotal importance to incubate in the presence of a COX-inhibitor to exclude in-vitro regulatory responses and subsequent metabolic changes. A significant change can be seen after 3 and 4 weeks of prostaglandin treatment, respectively. While in control vascular tissue the arachidonic acid conversion to 6-oxo-$PGF_{1\alpha}$, the stable derivative of PGI_2, is the highest (Tables 6, 7), the conversion rate is significantly depressed after administration of either of the PG's indicating a down-regulatory role. In contrast, however, the conversion rate to 6-oxo-$PGF_{1\alpha}$ is not affected in contractile SMC. These findings can be obtained at comparable extent in human femoral and popliteal arteries. Determination of the biologically active compound PGI_2 results in similar findings indicating that exogenous prostaglandin administration decreases PGI_2-production of ASMC, while the synthetic capacity of the contractile type apparently is not affected at all. All these findings are comparable for the femoral and the popliteal artery. The difference in response to PGE_1 and PGI_2 between animal experiments and human may well be due to the fact, that in various animal species it was quite difficult to induce receptor desensitization (33) as compared to humans. This problem was first coming up, when the „rebound" receptor desensitization on platelets of patients being infused continuously with PGI_2 was discovered by us in 1981. All together, these findings show a large number of benefits induced by PG-therapy on SMC functional behaviour. In particular, they indicate that the rebound activation (desensitization at the receptor level) may be of key importance not only regulating platelet activity and vascular thrombogenicity, but also SMC-proliferative response (30) and production of extracellular matrix.

Conclusions

PG-therapy decreases significantly mitotic activity as determined by $[^3H]$-thymidine incorporation. The number of ASMC in the plaque and the media underneath is decreased, as is the uptake of $[^{35}S]$ and $[^{14}C]$-proline, as indicators of glycosaminoglycan- and collagen-synthesis by vascular SMC. The extent of all these effects is clearly associated with the duration of PG-therapy and PGE_1 at the longer treatment periods tends to be more effective. AA conversion to PGI_2 is significantly enhanced in ASMC. This is apparently due to activation for the gene of COX-2. PGE_1 and PGI_2 in the vessel wall are working via different receptors. PG-therapy abolishes this increase in ASMC, while CSMC conversion is unaffected. Data on PG binding of SMC are underway at present, preliminary findings indicating that desensitization after PGI_2 (no data available so far for iloprost) but not after PGE_1 may occur. This knowledge and these benefits may gain even more relevance when analogues for a long-term (oral) PG-therapy will enter clinical routine.

Acknowledgements
The valuable help by W. Firbas, J. O'Grady, B.A. Peskar, Waltraud Rogatti, O. Wagner and T. Zidek, as well as the technicians Marianne Freudmann, Susanne Granegger and Sonja Reiter, is gratefully acknowledged.

References

1.   Pietila K, Moilanen T, Nikkari T. Prostaglandins enhance the synthesis of glycosaminoglycans and inhibit the growth of rabbit aortic smooth muscle cells in culture. Artery 1980; 7:509-518.

2.   Morita I. Regulation of cell growth by prostaglandins. Gan-To-Kagaku-Ryoho 1983; 10:1919-1929.

3.   Hüttner II, Gwebu ET, Panganamala RV, Sharma HM, Geer JC. Fatty acids and their prostaglandin derivatives: inhibitors of proliferation in aortic smooth muscle cells. Science 1977; 197:289-291.

4.   Smith DL, Willis AL, Mahmud I. Eicosanoid effects on cell proliferation in vitro: relevance to atherosclerosis. Prostaglandins Leukotr Med 1984; 16:1-10.

5.   Sinzinger H. Inhibition of mitotic and and proliferative activity of smooth muscle cells by prostaglandin E_1. In: Prostaglandin E_1 in Atherosclerosis. Sinzinger H, Rogatti W, editors. Berlin-Heidelberg-New York: Springer-Verlag, 1986: 39-48.

6.   Nilsson J, Olsson AG. Prostaglandin E_1 inhibits DNA synthesis in arterial smooth muscle cells stimulated with platelet-derived growth factor. Atherosclerosis 1984; 53:77-82.

7.   Sinzinger H, Kefalides A, Hoche C. Is the platelet derived growth factor (PDGF) a main regulator in atherosclerosis? Circulation 1982; 66:192.

8.   Sinzinger H, Steurer G, Kaliman J, Ettl K. The interaction between the platelet derived growth factor (PDGF) and prostaglandin (PG) I_2 is important for atherosclerosis. Adv Prostaglandins Thrombox Leukotr Res 1987; 17:216-218.

9.   Sinzinger H, Zidek T, Fitscha P, Kaliman J, Steurer G. Platelet derived growth factor (PDGF) and prostaglandins (PGE_1, PGI_2) as modulators of the atherogenetic process. Folia Haematol 1988; 115:439-442.

10.   Loesberg C, Vanwijk R, Zandbergen J, Vanagen WG, Vanmourik JA, Degroot PG. Cell cycle-dependent inhibition of human vascular smooth muscle cell proliferation by prostaglandin E_2. Exp Cell Res 1985; 160:117-125.

11.   Silberbauer K, Sinzinger H, Winter M. Prostacyclin production by vascular smooth muscle cells. Lancet 1978; i:1356-1357.

12.   Sinzinger H, Silberbauer K, Auerswald W. Prostacyclin production by vascular smooth muscle and endothelial cells. Atherosclerosis 1980; V:140-143.

13.   Larrue J, Bricaud H, Sinzinger H. Prostacyclin synthesis by proliferative aortic smooth muscle cells. VASA 1984; 13:311-318.

14. Owen NE. Effect of prostaglandin E_1 on DNA synthesis in vascular smooth muscle cells. Am J Physiol 1986; 250:C584-C588.

15. Zidek T, Steurer G, Fitscha P, Sinzinger H. Beneficial effect of prostaglandin I_2 on smooth muscle cell proliferation. In: Prostaglandins in Clinical Research. Sinzinger H, Schrör K, editors. Philadelphia-New York: Alan R. Liss Inc, 1987: 357-363.

16. Rimarachin JA, Jacobson JA, Szabo P, MacLouf J, Creminon C, Weksler BB. Regulation of cyclooxygenase-2 expression in aortic smooth muscle cells. Arterioscl Thromb 1994; 7:1021-1031.

17. Sinzinger H, Zidek T, Fitscha P, O'Grady J, Wagner O, Kaliman J. Prostaglandin I_2 reduces activation of human arterial smooth muscle cells in vivo. Prostaglandins 1987; 33:915-918.

18. Sinzinger H, Fitscha P, Wagner O, Kaliman J, Rogatti W. Prostaglandin E_1 decreases activation of arterial smooth muscle cells. Lancet 1986; i:156-157.

19. Feigl W, Sinzinger H, Wagner O, Leithner C. Quantitative morphological investigations on smooth muscle cells in vascular surgical specimens and their clinical importance. Experientia 1975; 31:1352-1353.

20. Moncada S, Gryglewski RJ, Bunting S, Vane JR. An enzyme isolated from arteries transforms prostaglandin endoperoxides to an unstable substance that inhibits platelet aggregation. Nature 1976; 263:663-665.

21. Ross R, Glomset JA. The pathogenesis of atherosclerosis. N Engl J Med 1976; 295:268-277.

22. Habenicht AJR, Goerig M, Grulich J, Rothe D, Gronwald R, Loth U, Schettler G, Kommerel B, Ross R. Human platelet-derived growth factor stimulates rapid prostaglandin synthesis by activation and by rapid de-novo synthesis of cyclooxygenase. J Clin Invest 1985; 75:1381-1387.

23. Lin AH, Bienkowski MJ, Gorman RR. Regulation of prostaglandin H synthase mRNA levels and prostaglandin biosynthesis by platelet derived growth factor. J Biol Chem 1989; 264:17379-17383.

24. Parischa PJ, Hassoun PM, Teufel E, Landman MJ, Fanbury BL. Prostaglandin E_1 and E_2 stimulate the proliferation of pulmonary artery smooth muscle cells. Prostaglandins 1992; 43:5-19.

25. Nakao Y, Ooyama T, Chang WC. Platelets stimulate aortic smooth muscle cell migration in vitro. Atherosclerosis 1982; 43:143-150.

26.   Yi Fan Y, Ramos KS, Chapkin RS. Cell cycle related inhibition of mouse vascular
       smooth muscle cell proliferation by prostaglandin E_1: relationship between
       prostaglandin E_1 and intracellular cAMP levels. Prostaglandins Leukot Essent Fatty
       Acids 1996; 54:101-107.

27.   Sinzinger H, Fitscha P, Zidek T, Firbas W. Beneficial effect of prostaglandin E_1 on
       smooth muscle cell proliferation. Prostaglandins Clin Res 1987; 2:351-355.

28.   Fitscha P, Rauscha F, Rogatti W, Peskar BA, O'Grady J, Sinzinger H. 13,14-dihydro-
       PGE_1, an in-vivo metabolite of PGE_1, decreases mitotic activity induced by
       corticosteroid administration. Eicosanoids 1991; 4:231-233.

29.   Sinzinger H, Zidek T, Rogatti W. PGE_1-pretreatment abolishes increased mitotic
       activity induced by stress. Exp Pathol 1988; 34:61-64.

30.   Grosser T, Bönisch D, Zucker T-P, Schrör K. The inhibition of growth factor-
       stimulated mitogenesis of coronary artery smooth muscle cells by prostacyclin is
       attenuated by homologous receptor desensitization. Circulation 1994; 90:I-636
       [abstract].

31.   Sinzinger H, Zidek T, Fitscha P, Wagner O, Rogatti W. PGI_2 and PGE_1 inhibit smooth
       muscle cell proliferation, mitotic activity and extracellular matrix formation. Thromb
       Haemost 1986; 21:999.

32.   Sinzinger H, Fitscha P, Wagner O, Kaliman J, Zidek T, Rogatti W. Additive benefit of
       PGI_2 and PGE_1 (via different mechanisms?) on inhibition of activation of human
       vascular smooth muscle cells? Exp Pathol 1990; 40:55-60.

33.   Moncada S. Personal communication 1985.

AAS 48
Prostaglandins and Control of Vascular
Smooth Muscle Cell Proliferation
© 1997 Birkhäuser Verlag Basel

Prostacyclin and nitric oxide-related gene transfer in preventing arterial thrombosis and restenosis

K.K. Wu

University of Texas-Houston Medical School, Vascular Biology Research Center and Division of Hematology, 6431 Fannin, Houston, Texas 77030, USA

Summary. Prostacyclin (PGI$_2$) and nitric oxide (NO) are potent vascular mediators, playing key roles in protecting arterial wall from injury-induced lesions. The key enzyme that catalyzes PGI$_2$ biosynthesis is cyclooxygenase (COX). COX-1 undergoes auto-inactivation, which severely limits PGI$_2$ synthesis. Overexpression of COX-1 in cultured endothelial cells by COX-1 gene transfer was accompanied by a higher capacity for and sustained synthesis of PGI$_2$. Adenovirus-mediated COX-1 gene transfer to angioplasty damaged carotid arteries in pigs augmented PGI$_2$ synthesis and prevents thrombus formation. Transfer of endothelial NO synthase (eNOS) into angioplasty injured, carotid arteries was reported to suppress intimal hyperplasia in rats. Transfer of PGI$_2$ and NO synthetic enzymes restores the vasoprotective properties and represents an exciting new strategy for treating arterial thrombotic disorders.

Introduction

Prostacyclin (PGI$_2$) and nitric oxide (NO) are potent vascular mediators that play a key role in vasoprotection (1, 2). They act synergistically on inhibiting platelet aggregation and adhesion (3). Both agents exhibit properties of vasodilation and blocking leukocyte adhesion to endothelial surface (3). Under physiological conditions, PGI$_2$ and NO are thought to be continuously synthesized and released by the endothelium to maintain arterial homeostasis. Under pathological stresses, their productions are enhanced through augmented de novo synthesis of their respective synthetic enzymes. Both PGI$_2$ and NO have very short biological half-life, and the key enzymes catalyzing their synthesis undergo autocatalysis and also have very short half-life. The cellular levels of these key synthetic enzymes, notably cyclooxygenase (COX) and nitric oxide synthase (NOS) are replenished by regulated gene transcription and enzyme synthesis. Genes coding for these enzymes are cloned and characterized (4). This has facilitated the investigations of regulation and augmentation of these genes.

PGI$_2$ biosynthesis

PGI$_2$ is synthesized primarily by vascular endothelial cells (EC) and smooth muscle cells (SMC). Its synthesis is catalyzed by three sequential enzymatic steps: (i) when cells are activated, the translocated, activated phospholipase A$_2$ (PLA$_2$) catalyzes the liberation of arachidonic acid (AA) from the sn-2 position of membrane phospholipids; (ii the released AA is converted to prostaglandin G$_2$ (PGG$_2$) and PGG$_2$ to PGH$_2$ by a bifunctional enzyme, cyclooxygenase (COX also known as prostaglandin H synthase or PGHS) and (iii) PGH$_2$ is converted to PGI$_2$ by prostacyclin synthase (PGIS). All three enzymes are constitutively expressed in EC and SMC. Phospholipase A$_2$ is a 84 kDa cytosolic protein, which is translocated to endoplasmic reticulum (ER) and nuclear membrane by a calcium-dependent mechanism. Two isoforms of COX have been characterized (5). COX-1 is constitutively expressed. It is a 70-kDa glycoprotein originally purified from ram seminal vesicle (6, 7). Human COX-1 gene is mapped to the long arm of chromosome 9. It spans 22 kb and contains

11 exons (8). The open reading frame of its transcript encodes a 599 amino acid polypeptide (9). The COX-1 enzyme is localized at the ER membrane via hydrophobic interactions. Recent studies have shown that COX-1 gene transcription in cultured EC is upregulated by phorbol 12-myristate 13-acetate (PMA) and interleukin-1 with a corresponding increase in COX-1 protein levels and enzyme activities (10). COX-2 cDNA was initially cloned from src-transformed chick embryo fibroblasts and PMA-induced murine 3T3 fibroblasts (11, 12). COX-2 cDNA was subsequently cloned from EC (13, 14). The human COX-2 cDNA shares about 60% sequence identity with COX-1 cDNA. It encodes a 602-amino acid polypeptide with a molecular weight of approximately 72 kDa (13, 14). It is not constitutively expressed but highly inducible. EC COX-2 has been reported to be inducible by IL-1 , PMA, serum, and lysophosphatidylcholine (4) and SMC COX-2 has been reported to be induced by serum (15). COX-2 gene spans ~8 kb on the long arm of chromosome 1 (16). The 5'-flanking regions (5'-FR) of COX-2 and COX-1 genes show distinct features. The COX-2 5'-FR bears a canonical TATA box and a multitude of regulatory elements characteristic of an immediate early gene while that of COX-1 gene has no TATA box and is G+C rich, characteristic of a housekeeping gene (8, 16). The minimal promoters of both genes have been delineated but the mechanisms by which the basal and induced transcription of these two genes operate, are not completely elucidated. Prostacyclin synthase (PGIS) catalyzes the final specific conversion of PGH_2 to PGI_2. PGIS cDNA cloned from bovine and human sources show an almost complete sequence identity between these two species (17). It encodes a 500-aa protein. Earlier studies suggest that PGIS is localized on the ER but in view of localization of cytosolic PLA_2 and COX-2 on the nuclear envelope, further studies are required to determine whether PGIS may be located at nuclear envelope or other sites besides ER.

Biosynthesis of nitric oxide by endothelial cells

Biosynthesis of NO is catalyzed by NO synthase (NOS). Three isoforms of NOS have been identified: (1) constitutive neuronal NOS (nNOS or NOS-I); (2) constitutive endothelial NOS (eNOS or NOS-III); and (3) inducible NOS (iNOS) or NOS-II) (for a review, please see ref. 4, 18, and 19). Cultured human EC expresses only eNOS. eNOS is membrane bound via myristoylation and palmitoylation. The constitutively expressed eNOS is inactive in quiescent cells. EC activation by agonists triggers intracellular calcium elevation which binds calmodulin (CaM) and the Ca/CaM complex binds to the CaM binding site of eNOS. This is considered to be an important biochemical step to facilitate electron transfer from the reductase domain located at the carboxyl half of eNOS molecule to the heme-containing region at the amino-half of the eNOS molecule where L-arginine is converted to L-citrulline and NO. eNOS is inactivated during catalysis. Human eNOS gene spans ~22 kb on the long arm of chromosome 7 q 35-36. It contains 26 exons and encodes a 1205-aa protein with a calculated molecular weight of about 133 kDa.

iNOS is expressed in human vascular smooth muscle cells. Human iNOS is homologous to the murine iNOS. It shows about 50-60% sequence identity with eNOS. Sequence comparison between these two isoforms of NOS shows conserved NADPH, FMN, and FAD binding sites at the reductase domain. The CaM binding site is also conserved but iNOS is catalytically active without intracellular calcium elevation. This is thought to be due to a high binding affinity of iNOS for Ca/CaM. Cultured vascular smooth muscle cell iNOS has been shown to be inducible by cyclic AMP analogs and other agonists. NO synthesis by SMC iNOS is about 3 orders of magnitude higher than that by EC eNOS and this large quantity of NO production is considered to be cytotoxic. In choosing the NOS isoform for gene transfer, eNOS gene transfer is considered to be more suitable for controlling the vascular disease processes without causing toxic effects on normal vascular cells.

COX and NOS induction by endothelial stress

Arterial endothelium is subjected to frequent stresses such as mechanical and shear stress, lipid and viral injuries. These injurious agents cause the endothelium to express leukocyte adhesive molecules and chemotactic compounds whereby inflammatory leukocytes transgress the endothelium and cause vascular inflammation and damage. These agents increase vascular permeability, express coagulant and antifibrinolytic molecules. We hypothesize that COX and NOS gene expressions are induced in response to these injurious agents to foster a strong protection of the arterial wall. To test this hypothesis, we chose to use lysophosphatidylcholine (lysoPC) as the model injurious agent. LysoPC is a major product of oxidized LDL and is considered to be a key mediator of oxidized LDL induced arterial wall injury and atherosclerosis (20, 21). Cultured human umbilical vein endothelial cells (HUVEC) (passage 1 or 2) were incubated in cultured medium containing 5% fetal bovine serum in the presence of lysoPC up to 100 μM. LysoPC at 100 μM or lower concentrations did not cause cell damage or morphological damage. At various time points, medium was removed and cells were washed. RNA was isolated. Northern blot analysis using COX-1 and COX-2 selective probes showed that LPC at 100 μM induced COX-2 mRNA expression but had no apparent effect on the basal level of COX-1 mRNA (22). Maximal levels of COX-2 mRNA induction were noted after the cells had been treated with lysoPC for 2-4 h (22). In parallel experiments, COX-1 and COX-2 protein levels were determined by Western blot analysis. The 72 kDa COX-2 protein band was significantly augmented by lysoPC at 100 μM and the time course and maximal level of protein increase were in accordance with those of mRNA induction (19). The PGI_2-synthetic activity of lysoPC treated HUVEC in response to arachidonate or ionophore A23187 stimulation was increased in accordance with the enzyme mass increase. The COX-1 protein levels, on the other hand, were unaltered by lysoPC. These results indicate that lysoPC differentially induces COX-2 expression in cultured HUVEC. We evaluated the effect of lysoPC on eNOS mRNA levels by Northern blot analysis. To our surprise, eNOS mRNA levels were increased by lysoPC in a concentration- and time-dependent manner (23). We did not anticipate this because eNOS like COX-1 is a "housekeeping gene" and had not been shown to be inducible. Increased eNOS mRNA levels were accompanied by increased eNOS protein levels. Based on densitometric estimates, maximal eNOS mRNA and protein levels induced by lysoPC were on average 5-fold higher than the basal eNOS mRNA and protein levels. The maximal enzyme activity on the other hand, was only about 2-fold over the basal enzyme activity. The reason for this is unclear. However, this level of augmented enzyme activity had an appreciable anti-platelet aggregation effect.

Taken together, our data indicate that in response to lysoPC stress, ECs increase the cellular levels of COX-2 and eNOS, allowing for a higher synthetic capacity for and a more sustained synthesis of these two vasoprotective molecules. We propose that COX-2 induction in EC serves as an important and powerful back-up mechanism for enhancing the vasoprotection. COX-2 induction in inflammatory tissues, on the other hand, is crucial in acute inflammation. Hence, the role of COX-2 induction depends on the tissues and the circumstances under which it is induced. Our study further shows that eNOS is inducible as well and the combined PGI_2 and NO augmented synthesis should have a beneficial effect on protecting the arterial wall from severe damage and probably on its reversal to normal.

When the insult such as oxLDL is severe and persistent, it may overwhelm the protective mechanism provided by stress-induced eNOS and COX-2 expressions and the endothelium become dysfunctional and eventually denuded. Persistent injuries by insulting agents are considered to be the major cause of atherosclerosis and other arterial lesions. This concept is illustrated in Figure 1.

We reasoned that at the early stage of severe arterial injury, transfer of COX and/or NOS may augment PGI_2 and NO synthesis, restore vasoprotective property and prevent thrombus, intimal hyperplasia and progression of atherosclerosis. This proposition is presented in Figure 2.

To test this hypothesis, we performed a series of experiments described below. The intimal issue was to determine whether COX-1 overexpression without concomitant PLA_2 and/or PGIS overexpressions would increase the capacity and duration of PGI_2 synthesis.

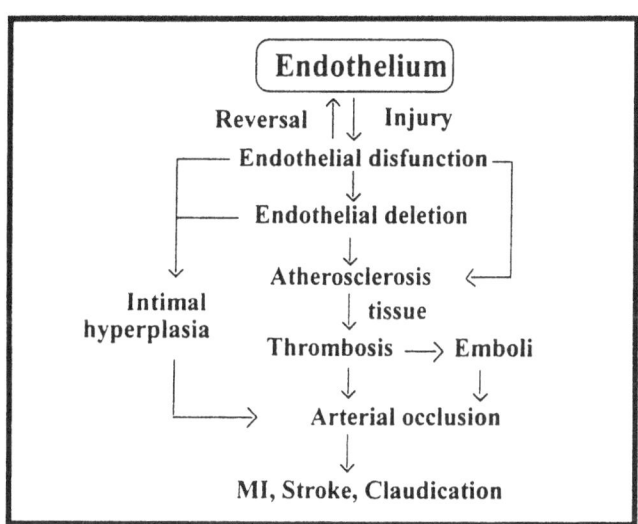

Figure 1. Reversal of endothelial damage. Endothelial injury is protected by vasoprotective molecules such as NO and PGI₂. However, when injury is severe, it may overwhelm the natural vasoprotective protective properties and lead to endothelial dysfunction, endothelial denudation with the eventual development of atherosclerosis, thrombosis and/or intimal hyperplasia.

Overexpression of COX-1 cDNA in cultured endothelial cells

COX-1 was considered to be the key step in determining the capacity of prostacyclin and other prostanoid synthesis primarily because of its autoinactivation during catalysis (24). To provide direct evidence to support this supposition, we stably transfected cultured human EC with a retroviral vector (BAG) containing a COX-1 cDNA insert (25). In the BAG vector, COX-1 cDNA was inserted into the BamHI sites in an antisense orientation relative to the 5'-LTR of the retroviral vector. COX-1 expression was driven by the late promoter of SV40 (25). The basal COX-1 mRNA and protein levels in these stably transfected cells are approximately 20-fold higher than those of the untransfected cells (25). PGI₂ synthesis in response to agonist stimulation in these stably transfected cells was increased by more than 20-fold over that of untransfected cells (25). Hence, these COX-1 overexpressed cells have an increased capacity for prostacyclin synthesis in response to physiological agonist.

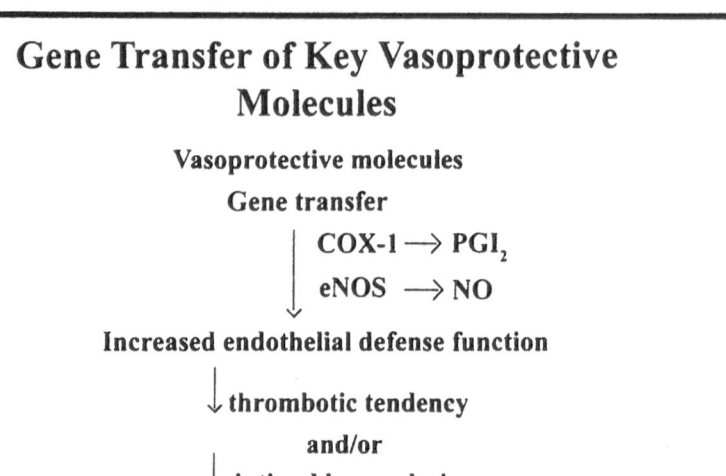

Figure 2. Concept of gene transfer for augmenting vascular synthesis of NO or PGI$_2$. In this scheme, we propose transfer of COX-1 and eNOS to increase the production of PGI$_2$ and NO, respectively.

The next question is: Will COX-1 overexpressed cells have a more sustained production of PGI$_2$ in response to repeated agonist challenge? It has been shown previously that PGI$_2$ biosynthesis by cultured EC or arterial tissues ex vivo in response to agonist stimulation was short lived followed by a refractory period (26, 27). The rate of PGI$_2$ production reached a plateau at 15-30 min after agonist treatment and declined rapidly thereafter. This has been attributed to suicidal autoinactivation of COX-1 during catalysis. In this COX-1 overexpressed EC, we provided a more quantitative analysis of PGI$_2$ synthetic kinetics during single and multiple episodes of agonist (ionophore A23187 or AA) challenge. During a single episode of challenge, our kinetic data indicate that the rate constant for decline of PGI$_2$ synthesis in response to AA or ionophore treatment was 0.064 ± 0.027 min^{-1} and 0.032 min^{-1}, respectively (25). The half-life of PGI$_2$ synthesis was calculated to be 11 min and 22 min, respectively (28). However, when cells were subjected to hourly AA or ionophore stimulation, the calculated half-life of prostacyclin synthesis was an order of magnitude longer than that of a single challenge (Table 1). Even after four hourly AA or ionophore stimulations, the cells still possess

COX-1 enzyme activity to catalyze the formation of an appreciable quantity of prostacyclin albeit only at about a 20% level of the initial capacity of PGI_2 synthesis. These results imply that in these COX-1 overexpressed cells, only a fraction of total COX-1 was autoinactivated during each hourly treatment of the cells with physiological agonists (28). The reduction of COX-1 protein masses estimated by Western blot analysis was compared with the reduction in PGI_2 formation during repeated hourly treatment with AA. The loss of protein mass is in agreement with the decline of PGI_2 synthesis (28). These in vitro data indicate that overexpression of COX-1 not only increases the capacity for PGI_2 synthesis but also prolongs the duration of PGI_2 production. They give credence for using COX-1 gene transfer in protecting against arterial damage.

Table 1. Kinetic analysis of decay of endothelial prostacyclin synthesis in response to single or multiple hourly treatments with AA, ionophore or PGH_2

	Single[§]			Multiple[†]		
	AA	Iono	PGH_2	AA	Iono	PGH_2
Rate Constant (min^{-1})	0.064[‡]	0.032	0.13	0.0078	0.0047	0.019
Maximal Capacity ($ng/10^6$ cells)	314	240	40			

[§] "Single" denotes incubation of COX-1 overexpressed EC with AA (10 μM), ionophore 23187 (10 μM) or PGH_2 (10 μM) and at various time points (0-240 min), medium was removed and its 6-keto-PGF_1 content was measured by RIA. The data fit a single exponential curve from which rate constant and maximal capacity of PGI_2 synthesis were determined.
[†] "Multiple" denotes treatment of COX-1 overexpressed EC with AA (10 μM), ionophore A23187 (10 μM) or PGH_2 (10 μM) for 1 hour. The medium was removed and the cells were incubated with fresh medium containing the same concentrations for 1 h. These hourly incubations were repeated for 5 hours. The hourly PGI_2 quantity was plotted and fit a single exponential curve wherein rate constant was determined.
[‡] All the values represent the mean value of several experiments (28).

PGIS was inactivated during catalysis (28). We have recently determined the kinetics of its inactivation by treating HUVEC with exogenously added PGH_2 (10 μM) and at various time points, the 6-keto-$PGF_{1\alpha}$ content released into the medium was measured by RIA. 6-keto-$PGF_{1\alpha}$ was rapidly formed and reached plateau at ~30 min. The kinetic curve fits a single exponential plot and the rate constant was determined as 0.13 ± 0.04 min^{-1}. The half-life of PGI$_2$ synthesis from PGH$_2$ was calculated to be <10 min. The maximal amount of PGI$_2$ produced from PGH$_2$ was only about 1/5 of that produced from AA (Table 1). When HUVECs were incubated hourly with PGH$_2$ and at the end of the hourly incubation, the medium was removed for 6-keto-PGF$_{1\alpha}$ determination and the cells were incubated in fresh medium for an hour. The 6-keto-PGF$_{1\alpha}$ content released into the medium declined with each hour of PGH$_2$ treatment. The results again show that not all enzymes were autoinactivated during the first hour of PGH$_2$ treatment. Decline of the hourly 6-keto-PGF$_{1\alpha}$ content with time follows a single exponential curve with a calculated half-life of about 60 min (Table 1) (28). This may be due to a limited substrate availability as the efficiency for the exogenously added PGH$_2$ to arrive at the substrate pocket of PGIS is a total mystery. However, substrate availability may not be the only mechanism. In this COX-1 overexpressed cell, stimulated with ionophore A23187 or exogenous AA, the capacity of PGI$_2$ synthesis is directly related to COX-1 levels in the cells and seems to be independent of PGIS levels. This would mean that PGIS in cultured human EC has a rapid turnover rate for catalyzing the conversion of PGH$_2$ to PGI$_2$. It is reasonable to assume that in our experimental model, only a fraction of PGIS is involved in catalyzing the conversion of PGH$_2$ to PGI$_2$. These kinetic data support the selection of COX-1 gene transfer to augment the capacity for PGI$_2$ synthesis and prolong the duration of PGI$_2$ synthesis in response to physiological and pathophysiological stimuli.

Augmentation of PGI$_2$ synthesis in damaged vessel wall by COX-1 gene transfer in vivo

Direct application of recombinant retroviruses to damaged arteries for overexpression of a transgene has been shown to be inadequate (29) because of two major drawbacks: (i) retroviruses infect only cells at active replication and (ii) relatively low titers of retroviruses (usually 10^4-10^6 pfu) are obtained. By contrast, direct application of adenoviral vectors containing a reporter gene insert has been shown to effectively express the reporter gene in the damaged arterial wall (30). Adenoviral vectors have major advantages over retroviruses: (i) much higher adenoviral titers can be prepared and (ii) adenoviruses infect both replicating and non-replicating cells. Since the damaged arterial wall is composed of non-replicating EC and replicating SMC, adenoviruses are expected to infect both types of cells and overexpress the gene products in these cells. It has been shown by Zoldhelyi et al (31) that replication-defective adenoviral vectors containing a COX-1 cDNA insert (Ad-COX-1) infected cultured human EC and porcine SMC and expressed COX-1 transgene mRNA. They further show that a 30 min incubation of EC with Ad-COX-1 was sufficient to produce significantly higher levels of COX-1 mRNAs and proteins than an identical treatment of cells with adenoviral vectors containing the E. coli. LacZ cDNA insert (Ad-LacZ) or buffer alone (31). These Ad-COX-1 infected cells produced a higher level of prostacyclin in response to AA or ionophore A23187 (28). These in vitro experiments indicate a high efficiency of COX-1 gene transfer conferred by adenoviral vectors.

The feasibility of direct instillation of Ad-COX-1 in COX-1 gene overexpression in vivo was evaluated in a pig carotid artery injury model of Butler et al (32). Recombinant Ad-COX-1 and its control vectors, Ad-LacZ and Ad-RR (without an insert of foreign gene) were prepared by homologous recombination (31). The pig carotid artery was isolated and subjected to injury with modified Spencer-Welles forceps. Ad-COX-1 (1×10^{10} pfu suspended in 0.5 ml buffer) was instilled in the injured arterial segment temporarily ligated for 30 min. The ligation was removed and flow reestablished. The animals were euthanized at 72 h after viral infection. Injured carotid arteries were removed and dissected into rings. The rings were incubated at 37°C in buffer containing 20 µM AA for 5 min and the 6-keto-PGF$_{1\alpha}$ contents in the media were measured by radioimmunoassay. When compared to the non-viral transfected injured carotid arteries, the Ad-COX-1 treated carotid arteries produced a three-fold increase in PGI$_2$. These results indicate that a 30 min treatment with Ad-COX-1 locally at the injured artery is sufficient for increasing COX-1 activity. However, a higher viral titer may be needed to achieve a higher COX-1 expression.

Prevention of thrombosis and intimal hyperplasia by COX-1 and eNOS gene transfer

The effect of Ad-COX-1 infection on preventing thrombosis as reported by Zoldhelyi et al (28) was evaluated in a porcine carotid angioplasty injury model modified from that described by Steele et al (33). In this model, complete arterial occlusion due to massive thrombus formation occurs frequently. A Fogarty balloon catheter was inserted into the carotid artery via a transfemoral route. After the catheter was in place, the balloon was inflated and the inflated catheter was pulled back and forth according to the procedure of Steele et al (33). The injured arterial segment was temporarily ligated and Ad-COX-1 or its vector controls was instilled in the segment for 30 min. The ligated regions were released to allow blood reflow. During the first 24 h of surgery, the animals were heparinized to prevent massive thrombosis. The blood flow of the injured carotid artery was monitored for 10 days after this initial period had become stabilized. At the end of day ten, the animals were euthanized and the carotid arteries were examined histologically for thrombus formation. In 8 animals, receiving a higher titer of Ad-COX-1 (3×10^{10} pfu in 0.5 ml of instilled buffer), complete occlusion of carotid arteries did not develop in any of the animals. Furthermore, all 8 animals maintained a normal carotid flow as detected by Doppler. By contrast, of 17 animals receiving a lower titer of Ad-COX-1 (0.5×10^{10} pfu in 0.5 ml infusate), 5 (30%) developed complete occlusion as evidenced by development of zero-flow on Doppler and large thrombus on histological examination (31). The results of the low titer Ad-COX-1 infection were not significantly different from those of the Ad-RR or the buffer control (31). The frequency of cyclic flow changes as detected by Doppler throughout the 10-day monitoring was also not statistically significantly different between the low-titer Ad-COX-1 group and the Ad-RR or the buffer control group. PGI_2 produced by injured carotid arteries infected with the high titer of Ad-COX-1 was approximately 5-fold higher than that produced by injured carotid arteries treated with viral vector control or buffer on day 10. PGI_2 produced by carotid arteries infected with the lower titer of Ad-COX-1 was not different from that produced by control arteries. These results suggest that COX-1 remained overexpressed in injured arteries receiving 3×10^{10} pfu Ad-COX-1. The viral titer appears important in providing a sustained expression of COX-1 and consequently conferring anti-thrombotic activities. In this animal model, there was minimal inflammatory cell infiltration in Ad-COX-1 infected tissues. Histological examinations suggested a reduction of intimal hyperplasia by Ad-COX-1 transfer but the results require further confirmation by carrying out longer experiments, i.e. 3-4 weeks to allow for a more quantitative evaluation of the intimal hyperplasia.

Nitric oxide (NO) has been shown to suppress SMC proliferation in in vitro cell culture experiments (3). When considering NO-related gene transfer for controlling intimal hyperplasia, eNOS gene transfer is a more logical selection. It has recently been reported by Von der Leyen et al (34) that transfer of eNOS into rat angioplasty-injured carotid arteries was accompanied by an augmented production of NO derivatives, nitrite and nitrate and a suppression of intimal hyperplasia. These results confirm the important role that NO plays in controlling smooth muscle proliferation and intimal hyperplasia.

Conclusion and prospective

Overexpression of COX-1 by virus-mediated gene transfer produces a higher capacity for and a more sustained synthesis of prostacyclin. Augmented PGI_2 synthesis in damaged arterial wall confers protection against thrombus formation and possibly also intimal hyperplasia. Transfer of eNOS cDNA into damaged arteries to augment the production of NO protects against intimal hyperplasia in a rat angioplasty model. These data provide strong evidence to support the concept that PGI_2 and NO play a key role in maintaining arterial homeostasis and protecting arteries against injury-induced pathological changes. Augmenting NO and PGI_2 productions locally at the arterial injured sites by gene transfer represents a new, exciting therapeutic strategy for several important human diseases, notably coronary heart disease and thrombotic stroke.

There are a number of issues that require resolution before gene transfer can be predictably and safely employed to treat human diseases. Major technical issues include: (1) development of efficient and non-toxic vectors and (2) development of targeted delivery of the desired gene to the injured sites. These two technical developments are especially critical when applying gene transfer to human atherosclerotic arteries. These complex lesions substantially hamper the entrance of vectors into the effector cells (35). Hence, high titer viral preparations without toxicity or high-efficient non-viral approaches must be developed in order to achieve sustained therapeutic effects.

Selection of vasoprotective and anti-proliferative genes is another major issue to be carefully considered. It is highly likely that a "combination gene therapy" approach may be better than a single gene therapy. One example of combination gene therapy is to construct hybrid vectors containing both COX-1 and eNOS cDNA inserts. This would allow for a concurrent augmentation of NO and PGI_2 synthesis. Given their synergistical activities, enhanced availability of NO and PGI_2 at the damaged arterial wall will not only prevent thrombosis and intimal hyperplasia but also other complex arterial lesions.

Acknowledgments
The author wishes to thank Xiao-Ming Xu, David Loose-Mitchell, Lee-Ho Wang, Pierre Zoldhelyi, James Willerson, Max Buja, Fred Clubb and Robert Meidell for collaboration. The work was supported by grants from National Institutes of Health (NS-23327 and HL-50675).

References

1.   Vane JR. Prostaglandins and the cardiovascular system. Br Heart J 1983; 49:405-409.

2.   Moncada S, Palmer RMJ, Higgs EA. Nitric oxide: Physiology, pathophysiology and pharmacology. Pharmacol Rev 1991; 43:109-142.

3.   Gryglewski RJ, Korbut R, Trabka-Janik E, Zembowicz A, Trybutec M. Interaction between NO donors and iloprost in human vascular smooth muscle, platelets, and leukocytes. J Cardiovasc Pharmacol 1989; 14(suppl 11):S124-128.

4.   Wu KK. Inducible cyclooxygenase and nitric oxide synthase. Adv Pharmacol 1995; 33:179-207.

5.   Smith WL, Marnett LJ. Prostaglandin endoperoxide synthase: Structure and catalysis. Biochim Biophys Acta 1990; 1083:1-17.

6.   Hemler M, Lands WEM, Smith WL. Purification of cyclooxygenase that forms prostaglandins. J Biol Chem 1976; 251:5575-5579.

7.   Miyamoto T, Ogino N, Yamamoto S, Hayaishi O. Purification of prostaglandin endoperoxide synthase from bovine vascular gland microsomes. J Biol Chem 1976; 251:2629-2636.

8.   Wang L-H, Hajibeigi A, Xu X-M, Loose-Mitchell D, Wu KK. Characterization of the promoter of human prostaglandin H synthase-1 gene. Biochem Biophys Res Comm 1993; 190:406-411.

9.   Funk CD, Funk LB, Kennedy ME, Pong AS, FitzGerald GA. Human platelet/erythroleukemia cell prostaglandin G/H synthase: cDNA cloning, expression, and gene chromosomal assignment. FASEB J 1991; 5:2304-2312.

10.   Xu X-M, Tang JL, Hajibeigi A, Loose-Mitchell DS, Wu KK. Transcriptional regulation of endothelial constitutive PGHS-1 expression by phorbol ester. Am J Physiol 1996; 270:C259-C264.

11.   Xie W, Chipman JG, Robertson DL, Erickson RL, Simmons DL. Expression of a mitogen-responsive gene encoding prostaglandin synthase is regulated by mRNA splicing. Proc Natl Acad Sci USA 1991; 88:2692-2696.

12.   Kujubu DA, Fletcher S, Varnum BC, Lim RW, Herschman HR. TIS10, a phorbol ester tumor promoter-inducible mRNA from Swiss 3T3 cells, encodes a novel prostaglandin synthase/cyclooxygenase homologue. J Biol Chem 1991; 266:12866-12872.

13.   Hla T, Nielson K. Human cyclooxygenase-2 cDNA. Proc Natl Acad Sci USA 1992; 89:7384-7388.

14.   Jones DA, Carlton DP, McIntyre TM, Zimmerman GA, Prescott SM. Molecular cloning of human prostaglandin endoperoxide synthase type II and demonstration of expression in response to cytokines. J Biol Chem 1993; 268:9049-9054.

15.   Pritchard KA Jr, O'Banion MK, Miano JM, Vlasic N, Bhatia UG, Young DA, Stemerman MB. Induction of cyclooxygenase-2 in rat vascular smooth muscle cells in vitro and in vivo. J Biol Chem 1994; 269:8504-8509.

16.   Tazawa R, Xu X-M, Wang L-H, Wu KK. Characterization of genomic structure, chromosome location and promoter of human prostaglandin H synthase-2 gene. Biochem Biophys Res Comm 1994; 203:190-199.

17.   Pereira B, Wu KK, Wang L-H. Molecular cloning and characterization of bovine prostacyclin synthase. Biochem Biophys Res Comm 1994; 203:59-66.

18.   Moncada S, Palmer RM, Higgs EA. Nitric oxide: physiology, pathophysiology and pharmacology. Pharmacol Rev 1991; 43:109-142.

19.   Marletta MA. Nitric oxide synthase structure and mechanism. J Biol Chem 1993; 268:12231-12234.

20.   Vidaver GA, Ting A, Lee JW. Evidence that lyso-lecithin is an important causal agent of atherosclerosis. J Theor Biol 1985; 115:27-41.

21.   Quinn MT, Parthasarathy S, Steinberg D. Lysophosphatidylcholine: A chemotactic factor for human monocytes and its potential role in atherosclerosis. Proc Natl Acad Sci USA 1988; 85:2805-2809.

22.   Zembowicz A, Jones SL, Wu KK. Induction of cyclooxygenase-2 in human umbilical vein endothelial cells by lysophosphatidylcholine. J Clin Invest 1995; 96:1688-1692.

23.   Zembowicz A, Tang J-L, Wu KK. Transcriptional induction of endothelial nitric oxide synthase type III by lysophosphatidylcholine. J Biol Chem 1995; 270:17006-17010.

24.   Egan RW, Paxton J, Kueh FA. Mechanism for irreversible self-deactivation of prostaglandin synthase. J Biol Chem 1976; 251: 7325-7335.

25.   Xu X-M, Ohashi K, Sanduja SK, Ruan K-H, Wang L-H, Wu KK. Enhanced prostacyclin synthesis in endothelial cells by retrovirus-mediated transfer of PGHS-1 cDNA. J Clin Invest 1993; 91:1843-1849.

26.   Kent RS, Diedrich SL, Whorton R. Regulation of vascular prostaglandin synthesis by metabolites of arachidonic acid in perfused rabbit aorta. J Clin Invest 1983; 72:455-465.

27.   McIntire TM, Zimmerman GA, Satoh K, Prescott SM. Cultured endothelial cells synthesize both PAF and PGI$_2$ in response to histamine, bradykinin and ATP. J Clin Invest 1985; 76:271-280.

28.   Sanduja SK, Tsai A-L, Matijevic-Aleksic N, Wu KK. Kinetics of prostacyclin synthesis in PGHS-1 overexpressed endothelial cells. Am J Physiol 1994; 267:C1459-1466.

29.   Flugelman MY, Jaklitsch MT, Newman KD, Casscells SW, Brattauer GL, Dichek DA. Low level in vivo gene transfer into the arterial wall through a perforated balloon catheter. Circulation 1991; 85:1110-1117.

30.   Lee SW, Trapnell BC, Rade JJ, Virmani R, Dichek DA. In vivo adenoviral vector-mediated gene transfer into balloon-injured rat carotid arteries. Circ Res 1993; 73:797-807.

31.   Zoldhelyi P, McNatt J, Xu X-M, Loose-Mitchell D, Meidell RS, Clubb FJ, Buja M, Willerson JT, Wu KK. Prevention of arterial thrombosis by adenovirus-mediated transfer of cyclooxygenase gene. Circulation 1996; 93:10-17.

32.   Butler KD, Ambler J, Dolan S, Giddings J, Talbot MD, Wallis RB. A non-occlusive model of arterial thrombus formation in the rat and its modification by inhibitors of platelet function, or thrombin activity. Blood Coagul Fibrinolysis 1992; 3:155-165.

33.   Steele PM, Chesebro JH, Stanson AW, Holmes DR, Dewanjee MK, Badimon K, Fuster V. Balloon angioplasty. Natural history of the pathophysiological response to injury in a pig model. Circ Res 1985; 57:105-112.

34.   Von der Lyen HE, Gibbons GH, Morishita R, Lewis NP, Zhang L, Nakajima M, Kaneda Y, Cooke JP, Dzau VJ. Gene therapy inhibiting neointimal vascular lesion: in vivo transfer of endothelial cell nitric oxide synthase gene. Proc Natl Acad Sci USA 1995; 92:1137-1141.

35.   Feldman LJ, Steg PG, Zheng LP, Chen D, Kearney M, McGarr SE, Bary JJ, Dedieu J-F, Perricaudet M, Isner JM. Low-efficiency of percutaneous adenovirus-mediated arterial gene transfer in the atherosclerotic rabbit. J Clin Invest 1995; 95:2662-2671.

List of contributors

G.K. Owens and G. Wise
Department of Molecular Physiology
and Biological Physics
University of Virginia
Health Sciences Center
Box 449, Jordan Hall
1300 Jefferson Park Ave.
Charlottesville, VA 22908-0001
U.S.A.

D. Praticò, M. Reilly, J.A. Lawson and
G.A. FitzGerald
Center for Experimental Therapeutics
and Clinical Research Center
Universität of Pennsylvania
School of Medicine
909 Biomedical Research Building
422 Curie Blvd.
Philadelphia, PA 19104
U.S.A.

G.W. Dorn II
College of Medicine
Department of Internal Medicine
Division of Cardiology
Universioty of Cincinnati
231 Bethesda Avenue
Cincinnati, OH 45267-0542
U.S.A.

K. Schrör, A.-A. Weber
Institut für Pharmakologie
Heinrich-Heine-Universität Düsseldorf
Moorenstr. 5
D-40225 Düsseldorf
Germany

H. Sinzinger, P. Fitscha, H. Kritz
Wilhelm Auerswald Atherosclerosis
Research Group (ASF)
Nadlergasse 1
and
Department of Nuclear Medicine
University of Vienna
Währinger Gürtel 18-20
A-1090 Vienna
Austria

K.K. Wu
Vascular Biology Research Center
and Division of Hematology
University of Texas - Houston
Medical School
6431 Fannin
Houston, TX 77030
U.S.A.

Subject index

Agents and Actions Supplements

Edited by **K. Brune**, *University of Erlangen, Germany* / **M.J. Parnham**, *Bonn, Germany*

Agents and Actions Supplements *(AAS)* is a book series for rapid publication of the proceedings of symposia on topics of current interest in inflammation, allergy, related respiratory diseases, thrombosis and related fields. The series allows fast dissemination of surveys and specialized reports on, for example, research into the role of prostaglandins in inflammation and thrombosis, new trends in the treatment of rheumatoid arthritis, allergic reactions and asthma.

N.S. Doherty, *Pfizer Central Research, Groton, CT, USA* / **B.M. Weichman,** *Wyeth-Ayerst Research, Princeton, NJ, USA* / **D.W. Morgan,** *Abbot Laboratories, Abbott Park, IL, USA* / **L.A. Marshall,** *SmithKline Beecham Pharmaceuticals, King of Prussia, PA, USA (Eds)*

W. Pruzanski / P. Vadas, *The Wellesley Hospital Research Institute, University of Toronto, Toronto, Ontario, Canada (Eds)*

K. Schrör, *Heinrich-Heine University, Düsseldorf, Germany* **C.R. Pace-Asciak,** *Hospital for Sick Children, Toronto, ONT, Canada (Eds)*

Inflammation: Mechanisms and Therapeutics

Novel Molecular Approaches to Anti-Inflammatory Therapy

Mediators in the Cardiovascular System: Regional Ischemia

All scientists involved in basic inflammation research in the pharmaceutical and biotechnology industries and academia will find this volume invaluable for their work. The book is primarily intended for pharmacologists, pathologists, immunologists, chemists, and rheumatologists.

1995. 224 pages. Hardcover
ISBN 3-7643-5129-2 (AAS 47)

Topics discussed include chemokines and their role in human disease, mediation of inflammation by cyclooxygenase-2, leukocyte adhesion and the anti-inflammatory effects of leukocyte integrin blockade, anti-inflammatory lipocortin-derived peptides, and the use of anti-PECAM and other agents in the control of inflammation.

1995. 192 pages. Hardcover
ISBN 3-7643-5096-2 (AAS 46)

Experts of international repute address in this volume relevant aspects of recent research and emerging themes in the field of myocardial ischemia. Containing the latest data in the field, this volume is of interest to pharmacologists, physicians, and all other scientists working in the fields of ischemia and mediator research.

1995. 332 pages. Hardcover
ISBN 3-7643-5130-6 (AAS 45)

Pharmacology • Rheumatology • Immunology

Birkhäuser Verlag • Basel • Boston • Berlin